BROOKINGS TRADE FORUM 2008/2009

Climate Change, Trade, and Competitiveness

Is a Collision Inevitable?

Lael Brainard

Isaac Sorkin

EDITORS

BROOKINGS INSTITUTION PRESS

Washington, D.C.

ISSN 1520-5479
ISBN 978-0-8157-0298-6

For information on subscriptions, standing orders, and individual copies, contact Brookings Institution Press, P.O. Box 465, Hanover, PA 17331-0465. Or call 866-698-0010. E-mail brookings@tsp.sheridan.com. Visit Brookings online at www.brookings.edu/press/bookstore.htm.

Brookings periodicals are available online through Online Computer Library Center (contact the OCLC subscriptions department at 800-848-5878, ext. 6251) and Project Muse (http://muse.jhu.edu).

BROOKINGS TRADE FORUM 2008/2009

Climate Change, Trade, and Competitiveness

Is a Collision Inevitable?

Foreword

This volume is the latest demonstration of the commitment by Brookings to contribute in every possible and appropriate way to finding a solution to the existential problem of climate change. The debate on this subject has shifted from science to public policy. Though no consensus has emerged, it is clear that addressing climate change effectively will require understanding the deep interactions between it and other policy areas. Reaching an international agreement for meaningful global action will require diplomacy at the highest level. Sustaining lower levels of greenhouse gas emissions will require a new energy infrastructure. And reducing emissions could end up reshaping almost every aspect of nations' domestic economies. Trade lies at the intersection of diplomacy and economic policy, and hence will be implicated in a push for action on climate change. *Climate Change, Trade, and Competitiveness: Is a Collision Inevitable?* examines this interaction from the perspective of economics, law, and international relations. The contributors discuss the role of trade in mitigating the negative effects of climate change on domestic industries, in determining the legality of climate change policy, and in reaching a global agreement on climate change. They lay out the complex decisions facing policymakers and make concrete suggestions for the path forward.

This volume, edited by Lael Brainard and Isaac Sorkin, includes papers by William Antholis of the Brookings Institution, Jason Bordoff of Brookings, Thomas Brewer of Georgetown University, Jeffrey Frankel of Harvard University, Warwick McKibbin of the Australian National University and Brookings, C. Ford Runge of the University of Minnesota, and Peter Wilcoxen of Brookings and Syracuse University. The volume includes comments by Joseph Aldy of Resources for the Future, Nils Axel Braathen of the Organization for Economic Cooperation and Development, Colin Bradford Jr. of Brookings, Daniel Drezner of Tufts University, (Tom) Hu Tao of the State Environmental Protection Administration of China, Arik Levinson of Georgetown University, Muthukumara Mani of the World Bank, and Andrew Shoyer of Sidley Austin LLP. The volume also includes concluding reflections from Jagdish

Bhagwati of Columbia University and the Council on Foreign Relations. The initial drafts of the papers, comments and reflections included in this volume were delivered at a conference held in Washington in June 2008.

The editors wish to thank Alfred Imhoff for rapid and precise copyediting, and Janet Walker of the Brookings Institution Press for her work in bringing the manuscript to publication. The authors remain responsible for the content of their chapters, including any errors or omissions. Special thanks are also due to Sandy Burke, Ann DeFabio Doyle, Kristie Latulippe, Anne Smith, and Amy Wong, who made the conference and the book possible.

This book, like the conference, was made possible by the generous support of the Doris Duke Charitable Foundation.

STROBE TALBOTT
President
Brookings Institution

Washington, D.C.
May 2009

LAEL BRAINARD
ISAAC SORKIN

Editors' Overview

Global climate change has leapt to the forefront of the public conscience. As policymakers grapple with the challenge of shaping policy to achieve a domestic and international consensus, attention has turned to the role that trade-related measures could play. In both Europe and the United States, policymakers have considered implementing so-called border adjustments on goods coming from countries with few or no climate change policies. These policymakers argue that such measures would protect domestic firms from "unfair" competition abroad and provide a stick to discipline laggard countries into implementing their own climate change policies.

Although this "negative" agenda of trade-as-stick has garnered most of the attention in policy debates, a "positive" agenda at the nexus of trade and climate change will be central in addressing the challenges posed by climate change. Trade and investment flows will help spread the technology necessary for climate change mitigation and adaptation and provide the financing for necessary investments. And as policymakers strive toward a post-2012 climate change framework, the success of the trading system in building a relatively successful international institution might provide lessons for the climate change system.

Because new climate change policies will likely be developed and implemented in the next few years, it is important to understand the relationship between climate change and the trading system, particularly whether it is desirable to include trade-related measures in climate change policies. To that end, practitioners, academics, and policymakers convened in Washington on June 9, 2008, to discuss these matters at a conference on climate change, trade, and competitiveness hosted by the Brookings Institution.

The chapters in this volume are revised versions of the papers presented at this conference. Each chapter is followed by one or more comments offered by the discussants at the conference. In chapter 1, Warwick McKibbin and Peter Wilcoxen examine the size and effects of border adjustments that would be needed to level the playing field if a carbon tax (or a carbon tax equivalent, like a cap-and-trade program) were put in place. Using their G-Cubed model of the world economy, they consider the effects of imposing carbon taxes both with and without border adjustments. In their model, the border adjustment is based on the carbon emissions associated with the production of each imported product, and the goal of the adjustment is to match the cost increase that would have occurred had the exporting country adopted a climate policy similar to that of the importing country. They analyze two cases. First, they examine what would happen if Europe imposed a carbon tax and then implemented a border adjustment based on U.S. energy intensities. In the second case, they examine what would happen if the United States imposed a carbon tax and then implemented a border adjustment based on Chinese energy intensities. In both cases, they find that the tariffs would be very small on most traded goods, would economically harm the countries imposing them, and would produce little in the way of environmental benefits.

In chapter 2, Jason Bordoff combines economic and legal analysis to weigh the expected benefits of border adjustments against their potential harms. Consistent with the analysis in chapter 1, he argues that a border adjustment on carbon-intensive imports from certain countries, such as that proposed in the Lieberman-Warner Climate Security Act, would do little to reduce the small amount of carbon leakage, although it would protect a few specific carbon-intensive domestic industries. He points out that there is also a risk that border adjustments would be abused for purely protectionist reasons, lead to retaliatory tit-for-tat trade wars, or be ruled noncompliant with the World Trade Organization (WTO). Though the outcome of any complex legal question is difficult to predict, Bordoff identifies several ways in which a border adjustment on carbon-intensive imports from countries without comparably effective climate policies could be inconsistent with WTO law. He also looks at the free allocation of allowances to compensate adversely affected industries and finds that such measures are more likely to be compliant with WTO law only to the extent that they are mostly ineffective in protecting employment and output in adversely affected industries. He concludes that the expected costs of both border adjustments and free allocation likely outweigh their benefits and suggests alternative mechanisms to address climate change while mitigating leakage and adverse effects on workers in carbon-intensive sectors.

In chapter 3, Jeffrey Frankel examines how to design border measures to minimize the risk of being inconsistent with WTO jurisprudence. He argues that if the measures are designed sensibly, there need not be a conflict between environmentalists who want climate change policies to include leveling mechanisms and free traders who want such policies to be nonprotectionist and consistent with WTO law. He points to two precedents—the shrimp-turtle case and the Montreal Protocol—that could justify border measures. On the basis of these, he suggests four principles for the design of border adjustments that would satisfy environmental and trade objectives. First, only countries participating in the Kyoto Protocol and/or its successors and following multilaterally agreed-on guidelines should use border measures, and then only against countries that are not participating in international agreements. Second, measures to address leakage to nonmembers should take the form of either tariffs or permit requirements on carbon-intensive imports. Third, independent panels of experts should be responsible for findings of fact, such as which countries are complying or not, which industries are involved and what is their carbon content, which countries are entitled to respond with border measures, and the nature of the response. Finally, import penalties should target fossil fuels and the roughly half dozen most-energy-intensive major industries—aluminum, cement, steel, paper, and glass, and perhaps iron and chemicals.

In chapter 4, Thomas Brewer advocates an expansion of the international negotiating agendas where climate change issues intersect trade and investment issues. He shows that although the current negotiating agenda emphasizes North-South technology transfers and financial flows, developing countries are sources as well as recipients of international technology transfers, and thus he concludes that the negotiating agenda should incorporate a more expansive geography of technology flows. Because private trade and investment flows are orders of magnitude larger than official development assistance and the main channel through which technology transfer occurs, he argues, the negotiating agenda should include a focus on the international institutional frameworks that affect trade and investment flows—especially foreign direct investment—by looking at the institutions and official barriers that inhibit them. This approach highlights the central role of multinational firms as both facilitators and inhibitors of technology transfers. As a result of his examination of the flows and the barriers that can impede them, Brewer suggests an expanded international negotiating agenda, including the joint agenda of the post-2012 climate change regime and the trade-investment regime.

In chapter 5, William Antholis takes up the question of what governance lessons can be learned from the trading system for the climate change system. He argues that the development of the General Agreement on Tariffs and Trade (GATT) / WTO system provides a road map for the development of confidence in a self-regulating system, which is what an effective climate change system requires. For him, the key features of the GATT/WTO system are that it built on a small *group* of states that, through a *general agreement*, were able to *gear up* domestic action over a *generation* and that also developed a mechanism to *graduate* nations when they emerge from the development process into the industrial world. He points out that the advantage of this five "Gs" approach is that it does not pose a direct challenge to national sovereignty. Instead, it coordinates the work of states in a way that respects the diversity of local governance and has a greater chance of gaining buy-in from the key players. He cautions that this approach does not guarantee fast domestic action, that many smaller states will feel left out of the process, and that the transition to the system may be difficult for many of these states.

In chapter 6, C. Ford Runge argues for the creation of a Global Environmental Organization (GEO). He points out that just as the GATT/WTO emerged from the postwar conferences as a rules-based response to increasing global commercial interdependence, multilateral responses to environmental challenges reflect a growing recognition of nations' pressing ecological interdependence, particularly with respect to global climate change. He argues that there is a substantial institutional gap in the ability of governments to respond to global environmental issues like climate change and to address trade-related environmental measures that can only be filled by a separate body like a GEO. A GEO could also disentangle trade from environmental matters, allowing the WTO to focus on the expansion of market access and reductions in trade protectionism. The WTO would need to pay attention to environmental measures only in cases of obvious trade distortion, whereas a GEO could help it clarify whether the environmental exceptions to the GATT articles are justified. And a critical factor for a GEO's success, Runge cautions, is that developing countries would need to be certain that it represents their interests.

Finally, in chapter 7, Jagdish Bhagwati reflects on the themes of the volume. He argues that unlike the GATT, which gave developing countries membership for free, no emissions reduction agreement would be effective without the full participation of developing countries. However, the stick of border adjustments is the wrong way to encourage developing country participation. Instead, the developed world should compensate the developing world for historic emissions in exchange for their participation in emissions reduction going forward. And regardless of the legal merits of border adjustments, Bhagwati recommends that the United States avoid popularizing them because of the chaos they might introduce into the international trading system.

WARWICK J. MCKIBBIN
PETER J. WILCOXEN

1

The Economic and Environmental Effects of Border Tax Adjustments for Climate Policy

For the foreseeable future, climate change policy will be considerably more stringent in some countries than in others. Indeed, the United Nations Framework Convention on Climate Change explicitly states that developed countries must take meaningful action before any obligations are to be placed on developing countries.

However, differences in climate policy will lead to differences in energy costs, and to concerns about competitive advantage. In high-cost countries, there will be political pressure to impose border tax adjustments (BTAs), or "green tariffs," on imports from countries with little or no climate policy and low energy costs. The BTAs would be based on the carbon emissions associated with the production of each imported product, and they would be intended to match the cost increase that would have occurred had the exporting country adopted a climate policy similar to that of the importing country.

Several justifications have been proposed for including BTAs as a key component of climate policy. Some researchers—including Stiglitz, Kopp and Pizer, and Ismer and Neuhoff—argue that BTAs are required for economic efficiency in carbon abatement.[1] An alternative argument is that BTAs are needed to keep climate policy from being undermined by the "leakage" of emissions through migration of carbon-intensive industries to low-tax countries and, as a corollary, to protect import-competing industries in high-tax countries.[2] There are also a number of papers that argue that the approach could be used to punish countries that did not participate in the Kyoto Protocol, or could be used as a threat to encourage recalcitrant countries to join a global regime.[3] Finally, there

The authors thank Nils Axel Braathen, Lael Brainard, Isaac Sorkin, and participants in the conference where this chapter was first presented for helpful comments.
1. See Stiglitz (2006); Kopp and Pizer (2007); Ismer and Neuhoff (2007).
2. For example, see Goh (2004); Hoerner (1998); Demailly and Quirion (2006); Babiker and Rutherford (2005).
3. For example, see Brack, Grubb, and Windram (2000); Hontelez (2007); and the discussion in Charnovitz (2003).

1

is also a considerable literature debating the legality of BTAs for climate polices under World Trade Organization rules.[4]

These arguments are reflected in the political debate in Europe and the United States. In 2006 then–French prime minister Dominique de Villepin suggested that countries that do not join a post-2012 international treaty on climate change should face additional tariffs on their industrial exports. The European Parliament's (2005/2049) resolution was focused on penalizing countries such as the United States for nonparticipation in the Kyoto Protocol. In the United States, both the Bingaman-Specter Bill (S 1766) and the Leiberman-Warner Bill (S 2191) include mechanisms that would, in effect, impose BTAs under some circumstances for imported goods from countries deemed to be making insufficient effort to reduce their greenhouse gas emissions.[5]

Most of the arguments in the literature, however, have been theoretical. Little empirical work has been done to determine either the magnitude that BTAs would take in practice, or on the economic and environmental consequences they would cause. This gap leads to a range of important questions. Would BTAs actually improve global carbon abatement? How much would they help or hurt the economy of the country imposing them? How much would they help or hurt the global economy? Are the gains, if any, large enough to justify the administrative costs involved? In this chapter, we address several of these questions. We estimate how large such tariffs would be in practice,[6] and then examine their economic and environmental effects using G-Cubed, a detailed multisector, multicountry model of the world economy.[7] We find that the tariffs would be small on most traded goods, would reduce leakage of emissions reduction very modestly, and would do little to protect import-competing industries. We conclude that the benefits produced by BTAs would be too small to justify their administrative complexity or their deleterious effects on international trade and the potentially damaging consequences for the robustness of the global trading system.[8]

In a sense, these results are not surprising, because most carbon emissions are from domestic activities, such as electricity generation and local and regional

4. See Biermann and Brohm (2005); Brewer (2008); Frankel (2005); Goh (2004); Hoerner (1998).

5. See the discussion in Brewer (2008).

6. This chapter focuses only on import adjustment. For a discussion of the problems that arise with adjustment to exports in order to maintain competitiveness, see Pearce and McKibbin (2007).

7. Other studies, such as Levinson and Taylor (2008), have used an econometric approach to examine a related issue, the "pollution haven hypothesis," to determine whether differences in historical environmental regulation have caused industries to migrate between countries. Our results, which examine prospective regulations using an econometrically estimated structural model and simulation analysis, produces results that are broadly consistent with that literature.

8. These results of the damaging effect on trade are also found in Droge and Kemfert (2005).

transportation, which are largely nontraded and are little affected by international trade.[9] In practice, the most important mechanism through which leakage could occur would be world oil markets, not trade in manufactured goods. A sufficiently large carbon tax imposed in a major economy would lower global oil prices and lead to higher consumption in countries with little or no carbon tax. However, BTAs would be neither appropriate nor effective in reducing that form of leakage. We conclude that it is an unnecessary distraction for the global community to focus much attention on negotiations over BTAs as a component of climate policy; they would not matter much in practice and, as also argued by Lockwood and Whalley, they may lead to greater distortions to the global trading system.[10]

An Overview of the G-Cubed Model

G-Cubed is an econometric intertemporal general equilibrium model of the world economy with regional disaggregation and sectoral detail. For this chapter, the world economy is divided into the ten regions shown in table 1-1. Each region is further decomposed into a household sector, a government sector, a financial sector, the twelve industrial sectors shown in table 1-2, and a capital-goods-producing sector. To facilitate the analysis of energy and environmental policy, five of the industries are used to represent segments of the energy industry: electric utilities, natural gas utilities, petroleum refining, coal mining, and crude oil and gas extraction. All regions are linked through bilateral trade in goods and financial assets. All relevant budget constraints are imposed on households, governments, and nations (the latter through accumulations of foreign debt). Households and firms have forward-looking expectations and use those projections when planning consumption and investment decisions. However, a portion of the households and firms are assumed to be liquidity constrained. G-Cubed is a very large example of the dynamic stochastic general equilibrium models used in the macroeconomics literature. It is also an intertemporal general equilibrium model from the computable general equilibrium class of models. We have described G-Cubed's theoretical and empirical structure in more detail elsewhere.[11] In the remainder of this section, we present a brief summary of its key features.

---

9. This point on the scale of leakage was made in McKibbin and Wilcoxen (1997).
10. Lockwood and Whalley (2008).
11. McKibbin and Wilcoxen (1998).

Table 1-1. Regions in the G-Cubed Model

1	United States
2	Japan
3	Australia
4	Europe
5	Other members of Organization for Economic Cooperation and Development (OECD)
6	China
7	India
8	Other developing countries (LDCs)
9	Eastern Europe and the former USSR (EEFSU)
10	Oil-exporting developing countries (members of the Organization of the Petroleum Exporting Countries, OPEC)

Table 1-2. Industrial Sectors in the G-Cubed Model

1	Electric utilities	7	Agriculture
2	Gas utilities	8	Forestry and wood products
3	Petroleum refining	9	Durable goods
4	Coal mining	10	Nondurables
5	Crude oil and gas extraction	11	Transportation
6	Other mining	12	Services

Producer Behavior

Each producing sector in each region is modeled by a representative firm, which chooses its inputs and its level of investment to maximize its stock market value subject to a multiple-input constant elasticity of substitution production function and a vector of prices it takes to be exogenous. We assume that output is produced using inputs of capital, labor, energy, and materials. Energy and materials, in turn, are aggregates of inputs of intermediate goods and services.

We assume that all regions share production methods that differ in first-order properties but have identical second-order characteristics. This is intermediate between the extremes of assuming that the regions share common technologies and of allowing the technologies to differ across regions in arbitrary ways.[12] Finally, the regions also differ in their endowments of primary factors and patterns of final demands.

Maximizing the firm's short-run profit subject to its capital stock and its production function gives the firm's factor demand equations. At this point, we

12. We adopt this approach because estimation of the second-order parameters requires a time series of input/output tables. Outside OECD countries there are generally too few tables available to permit the coefficients to be estimated separately for each country.

add two further levels of detail: We assume that domestic and imported inputs of a given commodity are imperfect substitutes, and that imported products from different countries are imperfect substitutes for each other. Thus, the final decision the firm must make is the fraction of each of its inputs to buy from each region in the model (including the firm's home country). We assume that all agents in each economy have identical preferences over foreign and domestic varieties of each particular commodity.[13] The result is a system of demand equations for domestic goods and imports from every region.

In addition to buying inputs and producing output, each sector must also choose its level of investment. We assume that capital is specific to each sector, that investment is subject to adjustment costs, and that firms choose their investment paths to maximize their market value. In addition, each industry faces the usual constraint on its accumulation of capital that the change in the capital stock is equal to gross investment less depreciation.

Following the cost of adjustment models of Lucas, Treadway, and Uzawa, we assume that the investment process is subject to rising marginal costs of installation.[14] Setting up and solving the firm's investment problem yields an investment decision that depends on production parameters, taxes, the current capital stock, and marginal q (that is, the ratio of the marginal value of a unit of capital to its purchase price).

Following Hayashi, we modify the investment function to improve its empirical properties by writing it as a function not only of q but also of the firm's current capital income.[15] This improves the empirical behavior of the specification and is consistent with the existence of firms that are unable to borrow and therefore invest purely out of retained earnings. The fraction of fully optimizing firms is taken to be 0.3 based on a range of empirical estimates;[16] the fraction that are liquidity constrained is 0.7.

In addition to the twelve industries discussed above, the model also includes a special sector that produces capital goods. This sector supplies the new investment goods demanded by other industries. Like other industries, the investment sector demands labor and capital services as well as intermediate inputs. We represent its behavior using a nested constant elasticity of substitution production function with the same structure as that used for the other sectors. However, we estimate the parameters of this function from price and quantity data for the final demand column for investment.

13. Anything else would require time-series data on imports of products from each country of origin to each industry, which is not only unavailable but difficult to imagine collecting.

14. Lucas (1967); Treadway (1969); Uzawa (1969).

15. Hayashi (1979).

16. These empirical estimates are reported by McKibbin and Sachs (1991).

Households and Governments

Households consume a basket of composite goods and services in every period and also demand labor and capital services. Household capital services consist of the service flows of consumer durables and residential housing. Households receive income by providing labor services to firms and the government, and from holding financial assets. In addition, they receive imputed income from ownership of durables and housing, and they also receive transfers from their region's government.

Within each region, we assume household behavior can be modeled by a representative agent who maximizes an intertemporal utility function subject to the constraint that the present value of consumption is equal to the sum of human wealth and initial financial assets. Human wealth is the present value of the future stream of after-tax labor income and transfer payments received by households. Financial wealth is the sum of real money balances, real government bonds in the hands of the public,[17] net holdings of claims against foreign residents, and the value of capital in each sector.

There has, however, been considerable debate about whether the actual behavior of aggregate consumption is consistent with the permanent income model.[18] On the basis of the evidence cited by Campbell and Mankiw,[19] we modify the basic household model described above to allow a portion of household consumption to depend entirely on current after-tax income (rather than on wealth). This could be interpreted in various ways, including the presence of liquidity-constrained households or households with myopic expectations. For the purposes of this chapter, we will not adopt any particular explanation and will simply take the income-driven share of consumption to be an exogenous constant. Following McKibbin and Sachs, we take the share to be 0.7 in all regions.[20]

Within each period, the household allocates expenditures among goods and services to maximize its intratemporal utility. In this version of the model, we assume that intratemporal utility may be represented by a Cobb-Douglas function of goods and services.[21] Finally, the supply of household capital services is determined by consumers themselves, who invest in household capital. We

17. Ricardian neutrality does not hold in this model because some consumers are liquidity-constrained.

18. Some of the key papers in this debate are Hall (1978); Flavin (1981); Hayashi (1982); and Campbell and Mankiw (1990).

19. Campbell and Mankiw (1990).

20. McKibbin and Sachs (1991). Our income-driven faction is somewhat higher than Campbell and Mankiw's estimate of 0.5.

21. This specification has the undesirable effect of imposing unitary income and price elasticities.

assume that households choose their level of investment to maximize the present value of future household capital service flows (taken to be proportional to the household capital stock), and that investment in household capital is subject to adjustment costs. In other words, the household investment decision is symmetrical with that of the firms.

Government

We take each region's real government spending on goods and services to be exogenous and assume that it is allocated among final goods, services, and labor in fixed proportions according to the base year input/output table for each region. Total government spending includes purchases of goods and services plus interest payments on government debt, investment tax credits, and transfers to households. Government revenue comes from sales, corporate, and personal income taxes and from the issuance of government debt. In addition, there can be taxes on externalities such as carbon dioxide emissions. We assume that agents will not hold government bonds unless they expect the bonds to be serviced. Accordingly, we impose a transversality condition on the accumulation of public debt in each region that has the effect of causing the stock of debt at each point in time to be equal to the present value of all future budget surpluses from that time forward. This condition alone, however, is insufficient to determine the time path of future surpluses: The government could pay off the debt by briefly raising taxes a lot; it could permanently raise taxes a small amount; or it could use some other policy. We assume that the government levies a lump sum tax in each period equal to the value of interest payments on the outstanding debt. In effect, therefore, any increase in government debt is financed by Consols (that is, bonds without a redemption date that pay interest in perpetuity), and future taxes are raised enough to accommodate the increased interest costs. Thus, any increase in the debt will be matched by an equal present value increase in future budget surpluses.

Macroeconomic Features:
Labor Market Equilibrium and Money Demand

We assume that labor is perfectly mobile among sectors within each region but is immobile between regions. Thus, within each region, wages will be equal across sectors. The nominal wage is assumed to adjust slowly according to an overlapping contracts model, where nominal wages are set based on current and expected inflation and on labor demand relative to labor supply. In the long run, labor supply is given by the exogenous rate of population growth; but in the short run, the hours worked can fluctuate depending on the demand for labor.

For a given nominal wage, the demand for labor will determine short-run unemployment.

Relative to other general equilibrium models, this specification is unusual in allowing for involuntary unemployment. We adopted this approach because we are particularly interested in the transition dynamics of the world economy. The alternative of assuming that all economies are always at full employment, which might be fine for a long-run model, is clearly inappropriate during the first few years after a shock.

Finally, because our wage equation depends on the rate of expected inflation, we need to include money demand and supply in the model. We assume that money demand arises from the need to carry out transactions and depends positively on aggregate output and negatively on the interest rate. The supply of money is determined by the balance sheet of the central bank and is exogenous.

International Trade and Asset Flows

The regions in the model are linked by flows of goods and assets. Each country's exports are differentiated from those of other countries; exports of durables from Japan, for example, are not perfect substitutes for exports of durables from Europe. Each region may import each of the twelve goods from potentially all the other regions. In terms of the way international trade data are often expressed, our model endogenously generates a set of twelve bilateral trade matrices, one for each good. The values in these matrices are determined by the import demands generated within each region.

Trade imbalances are financed by flows of assets between countries. We assume that asset markets are perfectly integrated across the regions and that financial capital is freely mobile.[22] Under this assumption, expected returns on loans denominated in the currencies of the various regions must be equalized period to period according to a set of interest arbitrage relations. In generating the baseline of the model, we allow for risk premiums on the assets of alternative currencies, although in counterfactual simulations of the model, these risk premiums are generally assumed to be constant and unaffected by the shocks we consider.

22. The mobility of international capital is a subject of considerable debate; see Gordon and Bovenberg (1996) or Feldstein and Horioka (1980). Also, this assumption should not be confused with our treatment of physical capital, which we assume to be specific to sectors and regions and hence completely immobile. The consequence of assuming mobile financial capital and immobile physical capital is that there can be windfall gains and losses to owners of physical capital. For example, if a shock adversely affects profits in a particular industry, the physical capital stock in that sector will initially be unaffected. Its value, however, will immediately drop by enough to bring the rate of return in that sector back into equilibrium with that in the rest of the economy.

For all regions other than China, we assume that exchange rates are free to float and that financial capital is freely mobile. This may appear less plausible for developing countries than it does for countries that belong to the Organization for Economic Cooperation and Development (OECD), because many developing countries have restrictions on short-term flows of financial capital. However, the capital flows in our model are the sum of short-term portfolio investment and foreign direct investment, and the latter is usually subject to fewer restrictions. In many countries with constraints on financial instruments, there are large flows of direct foreign investment responding to changes in expected rates of return. We assume that China pegs its exchange rate to the dollar, with a slight adjustment for deviations in output growth from trend and actual inflation from the desired target rate. This is closer to the recent historical record than the alternative assumptions of floating exchange rates or exactly fixed exchange rates.

Calculating the Carbon Content of Traded Goods

In general, BTAs are used to compensate for differences between countries in the taxes levied on goods, such as excise taxes or value-added taxes. Exporting countries may exempt traded goods from such taxes, or rebate taxes already collected, and importing countries may impose taxes equivalent to what would have been charged had the product been produced domestically. In this chapter, we examine only adjustments on imports and assume that carbon taxes are not rebated on exports. However, our methodology could be applied to export rebates as well.

The first step in computing a carbon-tax BTA on a given import would be to determine the total amount of fossil energy that was used directly or indirectly in the production of the good. Measuring direct energy consumption is relatively straightforward; an aircraft, for example, requires the direct use of energy when it is assembled. However, energy is also used indirectly through the production of all the parts and materials from which the plane is made. Computing total indirect energy consumption requires following the value-added chain back through intermediate products at every stage: Energy is used to produce sheet metal from aluminum; to produce aluminum from bauxite; and to mine the bauxite itself.

Tracing energy consumption all the way back to raw materials is possible using input/output tables. An input/output "use" table is a matrix showing the flow of each good to each industry in a particular year. Using that information, it is possible to determine the amount of each input needed to make a single unit of output. If A is a matrix of such coefficients, with one row for each input and one column for each output, the set of market equilibria for the inputs can be summarized in equation 1, where X is a vector of gross outputs by commodity, and F is a vector of final demands:

$$AX + F = X \tag{1}$$

The left side is total demand for each product: AX is the demand for intermediate goods, and F is final demand. The right side, X, is the supply of each good. Solving for X gives the total input needed to produce any given final demand vector:

$$X = (I - A)^{-1}F \tag{2}$$

Matrix $(I - A)^{-1}$ is known as a "total requirements" table. Each row corresponds to an input and each column to an output, and each element shows the amount of the input used directly or indirectly in the production of one unit of the output. For example, the total amount of coal consumption that can be attributed to production of a durable good would appear as an element in the coal row and durable goods column of $(I - A)^{-1}$.

Computing the implicit carbon content of each product requires two additional steps: The inputs of each fossil fuel are multiplied by appropriate emissions coefficients to convert fuel consumption to carbon emissions, and then carbon emissions are summed across fuels. The result is a single coefficient for each good giving the total carbon emissions that can be attributed to the good's production.

Because input/output tables are used in the construction of G-Cubed, the information needed to compute a total requirements table for each region in the model was readily available. In addition, the model's database includes emissions coefficients for each fuel, with emissions in millions of metric tons of carbon for each of the model's units of fuel, so the final steps were straightforward as well. Carrying out the calculation produced the results are shown in table 1-3. For convenience, the results are shown as thousands of metric tons. As indicated in the lower rows of the table, production of nonfuel traded goods generally involves emissions of 0.1 to 1.1 thousand metric tons per model unit of output. (The model's output units are large, corresponding to billions of dollars of output in a base year.) For example, one unit of durable goods produced in the United States is associated with 0.13 thousand metric tons of carbon. Implicit emissions vary strongly across regions; emissions associated with durables are only 0.7 thousand metric tons per unit in Japan, but are 1.01 thousand tons per unit in China. As expected, Japan and Europe are most efficient in terms of carbon and have the lowest coefficients; the highest coefficients are associated with China, India, and Eastern Europe and the former Soviet Union.

Table 1-3. Carbon Content of Nonfuel Exports by Country or Region of Origin

Thousands of metric tons of carbon per model unit

Export Sector	United States	Japan	Australia	Europe	Other OECD Members	China	India	Less-Developed Countries	Former USSR	OPEC
Electric utilities	2.65	0.38	2.76	0.60	1.45	7.63	4.98	2.07	4.27	1.05
Gas utilities	0.41	0.65	1.07	0.13	0.70	11.68	0.37	1.55	1.25	0.17
Petroleum refining	6.59	1.75	3.53	1.75	4.37	7.38	4.94	5.30	6.82	2.45
Coal	N.A.	N.A.	N.A.	N.A.	N.A.	N.A.	N.A.	N.A.	N.A.	N.A.
Crude oil	N.A.	N.A.	N.A.	N.A.	N.A.	N.A.	N.A.	N.A.	N.A.	N.A.
Mining	0.27	0.10	0.35	0.21	0.87	0.81	1.20	0.41	0.98	0.15
Agriculture	0.17	0.10	0.16	0.13	0.25	0.47	0.36	0.20	0.84	0.07
Forestry and wood	0.13	0.05	0.21	0.08	0.18	0.61	0.24	0.16	1.01	0.08
Durables	0.13	0.07	0.43	0.09	0.23	0.97	1.01	0.33	1.10	0.21
Nondurables	0.23	0.13	0.23	0.15	0.33	0.92	0.81	0.37	1.06	0.21
Transportation	0.22	0.08	0.25	0.18	0.32	0.87	0.59	0.35	1.08	0.20
Services	0.05	0.04	0.09	0.03	0.11	0.59	0.30	0.13	0.71	0.08

Source: Authors' calculations.
Note: OECD = Organization for Economic Cooperation and Development; OPEC = Organization of the Petroleum Exporting Countries; N.A. = not applicable.

Table 1-4. Carbon Tax and Border Tax Adjustment Simulations

Name	Description
EU-Tax	European carbon tax without BTAs
EU-TaxAdj	European integrated carbon tax and BTA policy
US-Tax	U.S. carbon tax without BTAs
US-TaxAdj	U.S. integrated carbon tax and BTA policy

Carbon Taxes with and without Border Tax Adjustments

This section describes the simulations we ran using the G-Cubed model to explore the effects of BTAs. We began by constructing a hypothetical carbon tax beginning at $20 per metric ton of carbon dioxide and rising by $0.50 per year to $40. The tax was intended to illustrate the effect of BTAs over a range of carbon prices but was not designed to achieve any specific emissions target. Our results would apply to a tradable permit policy as well, if the policy had similar equilibrium permit prices. However, administering the BTAs would be much more difficult under a permit system, because frequent revisions would be needed to follow fluctuations in the permit price.

We then examined the effects of the carbon tax under four scenarios about its implementation: (1) It is adopted in Europe without BTAs (referred to in the tables below as "EU-Tax"); (2) it is adopted in Europe, and BTAs are imposed on imports to Europe, assuming that the carbon embodied in the imports matches the energy intensity of the United States ("EU-TaxAdj"); (3) the tax is adopted in the United States without BTAs ("US-Tax"); and (4) it is adopted in the United States, and BTAs are imposed based on the energy intensity of China ("US-TaxAdj"). These simulations, which are descriptively summarized in table 1-4, were chosen to contrast the effects of BTAs between countries with similar and relatively efficient technology, Europe and the United States, with the effects of BTAs between countries with more heterogeneous technology, the United States and China.

In all four simulations, additional government revenue generated by the BTAs and the carbon tax itself was used to finance additional government spending in the corresponding region (that is, each region's fiscal deficit was held constant). Other fiscal assumptions could be used instead; for example, the revenue could be used to lower the deficit, or it could be returned to households via a lump-sum rebate.

The BTAs were computed by multiplying the embodied carbon per unit of output by the carbon tax prevailing in each year, and then converting the result to an ad valorem rate.[23] No adjustments were applied to imports of coal and crude petroleum, which are already subject to the carbon tax, which was applied to imports as well as domestic production. The results are shown in tables 1-5 and 1-6 for two carbon tax rates: $20 and $40 per ton. For the European tariffs shown in table 1-5, the rates for the $20 tax are small: less than 1 percent for tradable goods other than fuels. The rates for the $40 tax are twice as large, but still small; the largest are the tax on nondurables, at 0.92 percent, and on transportation, at 0.88 percent. For the U.S. tariffs shown in table 1-6, the rates are considerably higher. When the carbon tax is $20 per ton, the effective tariffs on durable and nondurable manufactured goods are almost 2 percent. At the $40 per ton rate, the tariffs double to slightly less than 4 percent. The rates in table 1-6 reflect the higher energy intensity of Chinese manufacturing, as was shown in table 1-3.

The effects of the two European scenarios on real gross domestic product (GDP) are shown in table 1-7. The carbon tax lowers European GDP by 0.6 to 0.7 percent. Lower European GDP, in turn, lowers GDP in Eastern Europe and the former Soviet Union (EEFSU) by 0.1 to 0.2 percent. OPEC's GDP also falls slightly, but the remaining countries and regions are affected by less than 0.1 percent. Adding BTAs has little additional effect on the European GDP, which is still reduced by 0.6 to 0.7 percent. However, the GDP of the EEFSU region drops considerably more than under the carbon tax alone: 0.5 to 0.7 percent. In part, this is due to the increase in trade barriers between Europe and the EEFSU; even though the BTA rates are calculated based on U.S. energy intensities, in this simulation they are applied to European imports.

The effects of the policies on annual carbon emissions from each region are shown in table 1-8. The carbon tax alone lowers European emissions by 53 to 98 million metric tons (mmt) per year over the 2010–30 period. Some of these emissions are offset by increases in other regions, often referred to as "leakage." In 2010, for example, European emissions fall by 53 mmt but world emissions fall by only 48 mmt. The difference is 5 mmt, or about 10 percent of the European decrease: 2 mmt in the United States, 1 mmt in developing countries, and 2 mmt in EEFSU. Adding BTAs causes a larger reduction in worldwide emissions: 69 to 127 mmt annually over the period. The larger cuts are the result of three interacting effects: European emissions do not fall as much (49 to 91 mmt), there is no leakage of emissions to the United States or less-developed countries, and the EEFSU's emissions fall by much more due to the much larger drop in the EEFSU's GDP.

---

23. The conversion to an ad valorem rate was for convenience; in practice, it is likely that a unit tax would be used.

Table 1-5. Simulated European Border Tax Adjustments Based on U.S. Energy Intensity

Percentage point change in ad valorem tariff

Sector	$20 per ton carbon tax	$40 per ton carbon tax
Electric utilities	5.30	10.60
Gas utilities	0.82	1.64
Petroleum refining	13.18	26.36
Coal	N.A.	N.A.
Crude oil	N.A.	N.A.
Mining	0.54	1.08
Agriculture	0.34	0.68
Forestry and wood	0.26	0.52
Durables	0.26	0.52
Nondurables	0.46	0.92
Transportation	0.44	0.88
Services	0.10	0.20

Source: Authors' calculations.
Note: N.A. = not applicable.

Table 1-6. Simulated U.S. Border Tax Adjustments Based on China's Energy Intensity

Percentage point change in ad valorem tariff

Sector	$20 per ton carbon tax	$40 per ton carbon tax
Electric utilities	15.26	30.52
Gas utilities	23.36	46.72
Petroleum refining	14.76	29.52
Coal	N.A.	N.A.
Crude oil	N.A.	N.A.
Mining	1.62	3.24
Agriculture	0.94	1.88
Forestry and wood	1.22	2.44
Durables	1.94	3.88
Nondurables	1.84	3.68
Transportation	1.74	3.48
Services	1.18	2.36

Source: Authors' calculations.
Note: N.A. = not applicable.

Table 1-9 shows the effects of the two policies on short-run interest rates in each region. Both policies lower the return to capital in Europe, and to a lesser extent, the EEFSU. The changes in interest rates in other regions are generally very small. Lower rates of return in Europe and the EEFSU lead to capital outflows and shifts of the two regions' trade and current account balances toward surplus, as shown in tables 1-10 and 1-11. The capital flows to the remaining regions in the model, which generally see their trade and current accounts shift

Table 1-7. Simulated Effects of European Policies on Real Gross Domestic Product, 2010, 2020, and 2030

Percentage changes from values for business as usual

Country or Group	EU-Tax			EU-TaxAdj		
	2010	2020	2030	2010	2020	2030
United States	0.0	0.0	0.0	0.0	0.0	0.0
Japan	0.0	0.0	0.0	0.0	0.0	0.0
Australia	0.0	0.0	0.0	0.0	0.0	0.0
Europe	−0.7	−0.6	−0.7	−0.7	−0.6	−0.7
Other OECD members	0.0	0.0	0.0	0.0	0.0	0.0
China	0.0	0.0	0.0	0.0	0.0	0.0
India	0.0	0.0	0.0	0.0	0.0	0.0
Less-developed countries	0.0	0.0	0.0	0.0	0.0	0.0
EEFSU	−0.2	−0.1	−0.1	−0.7	−0.5	−0.5
OPEC	−0.1	−0.1	−0.1	−0.2	−0.2	−0.2

Source: Authors' simulations.
Note: OECD = Organization for Economic Cooperation and Development; EEFSU = Eastern Europe and the former USSR; OPEC = Organization of the Petroleum Exporting Countries.

Table 1-8. Simulated Effects of European Policies on Carbon Emissions, 2010, 2020, and 2030

Millions of metric tons

Country or Group	EU-Tax			EU-TaxAdj		
	2010	2020	2030	2010	2020	2030
United States	2	2	2	0	0	0
Japan	0	0	0	0	0	0
Australia	0	0	0	0	0	0
Europe	−53	−72	−98	−49	−66	−91
Other OECD members	0	0	0	0	0	0
China	0	0	0	0	−1	−1
India	0	0	0	0	0	0
Less-developed countries	1	2	2	−1	−1	−1
EEFSU	2	3	5	−18	−24	−32
OPEC	0	0	0	−1	−2	−2
Total	−48	−64	−88	−69	−93	−127

Source: Authors' simulations.
Note: OECD = Organization for Economic Cooperation and Development; EEFSU = Eastern Europe and the former USSR; OPEC = Organization of the Petroleum Exporting Countries.

toward deficit. The euro strengthens slightly relative to the dollar, as shown in table 1-12. Exchange rates in the model are dollars per unit of foreign currency. An appreciation of the euro relative to the dollar, therefore, appears in the table as a percentage increase in the exchange rate for Europe.

The effects of the policies on European prices and domestic output are shown in table 1-13. The carbon tax policy, shown in the top section of the

Table 1-9. Simulated Effects of European Policies on Short-Run Interest Rates, 2010, 2020, and 2030

Percentage point change

Country or Group	EU-Tax			EU-TaxAdj		
	2010	2020	2030	2010	2020	2030
United States	−0.01	−0.01	−0.01	−0.01	−0.01	−0.01
Japan	−0.01	−0.01	−0.01	−0.02	−0.01	−0.01
Australia	−0.01	−0.01	−0.01	−0.02	−0.01	−0.01
Europe	−0.04	−0.03	−0.04	−0.05	−0.04	−0.05
Other OECD members	−0.01	−0.01	−0.01	−0.02	−0.01	−0.02
China	−0.02	−0.01	−0.01	−0.02	−0.01	−0.01
India	−0.01	−0.01	−0.01	−0.01	−0.01	−0.01
Less-developed countries	−0.02	−0.01	−0.01	−0.02	−0.01	−0.02
EEFSU	−0.03	−0.01	−0.02	−0.03	−0.02	−0.02
OPEC	−0.01	−0.01	−0.01	−0.01	−0.01	−0.01

Source: Authors' simulations.
Note: OECD = Organization for Economic Cooperation and Development; EEFSU = Eastern Europe and the former USSR; OPEC = Organization of the Petroleum Exporting Countries.

Table 1-10. Simulated Effects of European Policies on Trade Balances, 2010, 2020, and 2030

Billions of dollars

Country or Group	EU-Tax			EU-TaxAdj		
	2010	2020	2030	2010	2020	2030
United States	−2.1	−0.6	0.1	−2.1	−0.4	0.6
Japan	−1.0	−0.3	−0.1	−1.2	−0.4	−0.2
Australia	0.0	0.0	0.0	0.0	0.0	0.0
Europe	5.5	1.6	0.4	4.7	1.9	1.4
Other OECD members	−0.1	0.0	0.0	−0.1	0.0	0.1
China	−0.5	−0.2	−0.1	−0.6	−0.2	−0.1
India	−0.2	−0.1	0.0	−0.2	−0.1	−0.1
Less-developed countries	−1.3	−0.2	0.0	−1.2	−0.3	0.0
EEFSU	0.0	0.1	0.0	1.3	0.5	−0.2
OPEC	0.0	0.0	0.1	0.1	0.0	0.1

Source: Authors' simulations.
Note: OECD = Organization for Economic Cooperation and Development; EEFSU = Eastern Europe and the former USSR; OPEC = Organization of the Petroleum Exporting Countries.

table, raises coal prices sharply: by 23 percent in 2010, rising to 33 percent in 2030. Coal output drops by 8 percent in 2010, rising to 13 percent in 2030. Other energy prices, including the natural gas price, rise as well, although by much smaller percentages: 5 to 6 percent for crude oil and refined petroleum, and 1 to 2 percent for electricity. The combined tax and BTA policy shown in the bottom of the table is very similar, but with slightly larger increases in most prices (due to the tariffs) and slightly smaller reductions in output (due to the

Table 1-11. Simulated Effects of European Policies on Current Accounts, 2010, 2020, and 2030

Billions of dollars

Country or Group	EU-Tax			EU-TaxAdj		
	2010	2020	2030	2010	2020	2030
United States	−3.0	−2.7	−3.4	−3.1	−2.9	−3.7
Japan	−1.6	−1.3	−1.7	−1.9	−1.7	−2.2
Australia	0.0	0.0	0.0	0.0	0.0	0.0
Europe	7.9	6.7	8.6	6.5	6.1	8.5
Other OECD members	−0.3	−0.3	−0.3	−0.3	−0.2	−0.3
China	−0.7	−0.6	−0.7	−0.9	−0.7	−1.0
India	−0.2	−0.2	−0.3	−0.3	−0.3	−0.4
Less-developed countries	−1.9	−1.4	−1.7	−1.7	−1.4	−1.8
EEFSU	0.0	0.2	0.1	1.9	1.9	2.0
OPEC	−0.1	−0.1	−0.1	0.1	0.1	0.0

Source: Authors' simulations.
Note: OECD = Organization for Economic Cooperation and Development; EEFSU = Eastern Europe and the former USSR; OPEC = Organization of the Petroleum Exporting Countries.

Table 1-12. Simulated Effects of European Policies on Real Exchange Rates, 2010, 2020, and 2030

Percentage changes from values for business as usual

Country or Group	EU-Tax			EU-TaxAdj		
	2010	2020	2030	2010	2020	2030
United States	—	—	—	—	—	—
Japan	−0.1	−0.1	−0.1	−0.1	0.0	0.0
Australia	−0.1	−0.1	−0.1	−0.1	−0.1	−0.1
Europe	0.5	0.7	1.0	0.9	1.2	1.5
Other OECD members	0.0	0.0	0.0	0.0	0.0	0.1
China	−0.1	−0.1	0.0	−0.1	−0.1	0.0
India	0.0	0.0	0.0	0.0	0.0	0.0
Less-developed countries	−0.1	0.0	0.0	−0.1	−0.1	0.0
EEFSU	−0.3	−0.2	−0.1	−0.9	−0.8	−0.7
OPEC	−0.2	−0.2	−0.2	−0.3	−0.3	−0.3

Source: Authors' simulations.
Note: Exchange rates are measured as dollars per unit of foreign currency. OECD = Organization for Economic Cooperation and Development; EEFSU = Eastern Europe and the former USSR; OPEC = Organization of the Petroleum Exporting Countries.

shift away from imports to domestic production). However, the protective effect of the adjustments for European producers is very small, typically raising output by only 0.1 percent relative to the carbon tax alone.

Tables 1-14 through 1-20 show the effects of the two U.S. policies on the same set of variables. In general, the effects of the carbon tax are similar in magnitude but with the United States and the region representing other OECD members (which includes Canada and Mexico) filling the roles of Europe and

Table 1-13. Simulated Effects of European Policies on European Prices and Output, 2010, 2020, and 2030

Percentage changes from values for business as usual

	Prices			Quantities		
Sector	*2010*	*2020*	*2030*	*2010*	*2020*	*2030*
EU-Tax						
Electric utilities	1.6	1.9	2.2	−1.0	−1.1	−1.3
Gas utilities	0.5	0.4	0.5	−1.3	−1.6	−1.9
Petroleum refining	4.5	5.1	6.1	−2.8	−3.0	−3.4
Coal	22.5	27.8	33.4	−7.5	−9.8	−13.1
Crude oil	4.9	5.6	6.7	−3.3	−4.0	−5.1
Mining	0.5	0.4	0.5	−1.0	−0.8	−0.9
Agriculture	0.4	0.3	0.4	−0.1	−0.1	−0.2
Forestry and wood products	0.3	0.2	0.2	−0.5	−0.4	−0.4
Durables	0.3	0.1	0.2	−1.1	−0.6	−0.7
Nondurables	0.4	0.4	0.4	−0.2	−0.1	−0.2
Transportation	0.5	0.5	0.6	−0.4	−0.4	−0.4
Services	0.3	0.2	0.2	0.0	0.1	0.1
EU-TaxAdj						
Electric utilities	1.6	1.9	2.2	−0.9	−1.0	−1.2
Gas utilities	0.5	0.4	0.5	−1.3	−1.6	−1.9
Petroleum refining	4.8	5.6	6.6	−2.4	−2.6	−3.0
Coal	22.3	27.5	33.1	−7.5	−9.8	−13.0
Crude oil	4.8	5.5	6.5	−3.1	−3.7	−4.8
Mining	0.5	0.5	0.6	−1.2	−0.9	−1.0
Agriculture	0.3	0.3	0.3	−0.2	−0.2	−0.2
Forestry and wood products	0.2	0.1	0.1	−0.5	−0.4	−0.4
Durables	0.2	0.1	0.1	−1.2	−0.7	−0.8
Nondurables	0.4	0.4	0.4	−0.2	−0.2	−0.2
Transportation	0.5	0.5	0.6	−0.5	−0.4	−0.5
Services	0.2	0.1	0.2	0.1	0.2	0.2

Source: Authors' simulations.

the EEFSU. Table 1-14 shows that the carbon tax reduces U.S. GDP by 0.6 to 0.7 percent and other OECD members' GDP by 0.3 to 0.4 percent. Adding BTAs has negligible effect on U.S. GDP but increases the effect on other OECD members' GDP to reductions of 0.8 to 0.1 percent. Also, additional regions are affected as well, particularly developing countries.

As shown in table 1-15, the carbon tax reduces U.S. carbon emissions by much more than it reduced European emissions: 303 to 577 mmt per year over the period 2010–30. As with the European case, the carbon tax alone leads to some leakage of emissions: World emissions fall by 293 to 554 mmt. Leakage, therefore, ranges from 10 to 23 mmt, or 3 to 4 percent of the U.S. reduction.

Table 1-14. Simulated Effects of U.S. Policies on Real Gross Domestic Product, 2010, 2020, and 2030

Percentage changes from values for business as usual

Country or Group	US-Tax			US-TaxAdj		
	2010	2020	2030	2010	2020	2030
United States	−0.6	−0.6	−0.7	−0.6	−0.6	−0.7
Japan	0.0	0.0	0.0	−0.1	−0.1	−0.1
Australia	0.0	0.0	0.0	−0.1	−0.1	0.0
Europe	0.0	0.0	0.0	−0.1	−0.1	−0.1
Other OECD members	−0.4	−0.3	−0.3	−1.0	−0.8	−0.8
China	0.0	0.0	0.0	0.0	0.0	0.0
India	0.0	0.0	0.0	−0.1	−0.1	−0.1
Less-developed countries	−0.2	−0.1	−0.1	−0.5	−0.2	−0.2
EEFSU	0.0	0.0	0.0	−0.1	−0.1	−0.1
OPEC	−0.4	−0.3	−0.3	−0.5	−0.4	−0.3

Source: Authors' simulations.
Note: Exchange rates are measured as dollars per unit of foreign currency. OECD = Organization for Economic Cooperation and Development; EEFSU = Eastern Europe and the former USSR; OPEC = Organization of the Petroleum Exporting Countries.

Table 1-15. Simulated Effects of U.S. Policies on Carbon Emissions, 2010, 2020, and 2030

Millions of metric tons

Country or Group	US-Tax			US-TaxAdj		
	2010	2020	2030	2010	2020	2030
United States	−303	−422	−577	−279	−390	−535
Japan	0	0	0	−1	−1	−1
Australia	0	0	0	0	0	0
Europe	1	2	2	−2	−3	−3
Other OECD members	3	4	6	−4	−5	−6
China	0	0	0	−1	−2	−2
India	0	0	0	−1	−1	−1
Less-developed countries	5	8	11	−6	−4	−5
EEFSU	1	1	2	−2	−2	−2
OPEC	0	0	1	−1	−1	−2
Total	−293	−405	−554	−297	−407	−558

Source: Authors' simulations.
Note: Exchange rates are measured as dollars per unit of foreign currency. OECD = Organization for Economic Cooperation and Development; EEFSU = Eastern Europe and the former USSR; OPEC = Organization of the Petroleum Exporting Countries.

As with the European simulations, adding BTAs causes the U.S. reduction to be smaller but causes larger drops in emissions outside the United States and results in slightly larger global reductions: 297 to 558 mmt annually over the period 2010–30.

The effects on short-run interest rates are shown in table 1-16, and the main effect is a small reduction in rates in the United States. Under the carbon tax,

Table 1-16. Simulated Effects of U.S. Policies on Short-Run Interest Rates, 2010, 2020, and 2030

Percentage point change

Country or Group	US-Tax			US-TaxAdj		
	2010	*2020*	*2030*	*2010*	*2020*	*2030*
United States	−0.02	−0.03	−0.03	−0.05	−0.04	−0.04
Japan	−0.01	0.00	0.00	−0.01	0.01	0.01
Australia	−0.02	−0.01	−0.01	−0.03	−0.01	−0.01
Europe	−0.01	0.00	0.00	−0.02	0.00	0.00
Other OECD members	−0.01	0.00	0.00	0.03	0.00	0.00
China	−0.01	0.00	0.00	0.00	0.01	0.01
India	0.00	0.00	0.00	0.01	0.01	0.01
Less-developed countries	−0.01	0.00	0.00	0.01	0.01	0.01
EEFSU	−0.01	0.00	0.00	−0.01	0.00	0.00
OPEC	0.01	0.01	0.01	0.02	0.01	0.01

Source: Authors' simulations.
Note: Exchange rates are measured as dollars per unit of foreign currency. OECD = Organization for Economic Cooperation and Development; EEFSU = Eastern Europe and the former USSR; OPEC = Organization of the Petroleum Exporting Countries.

Table 1-17. Simulated Effects of U.S. Policies on Trade Balances, 2010, 2020, and 2030

Billions of dollars

Country or Group	US-Tax			US-TaxAdj		
	2010	*2020*	*2030*	*2010*	*2020*	*2030*
United States	−0.5	−0.9	0.4	−4.6	−1.2	4.3
Japan	−0.2	0.7	0.9	0.4	1.6	1.6
Australia	0.2	0.2	0.1	0.6	0.4	0.3
Europe	−1.0	−0.4	−0.8	0.5	−0.5	−2.7
Other OECD members	1.2	0.9	0.5	2.9	1.9	1.0
China	−0.4	0.0	0.0	−0.8	0.0	−0.3
India	0.0	0.1	0.2	0.1	0.3	0.3
Less-developed countries	0.3	−0.1	−0.4	1.4	−0.8	−1.6
EEFSU	0.4	0.5	0.5	1.0	1.1	1.0
OPEC	0.5	−0.3	−0.7	−0.1	−1.0	−1.5

Source: Authors' simulations.
Note: Exchange rates are measured as dollars per unit of foreign currency. OECD = Organization for Economic Cooperation and Development; EEFSU = Eastern Europe and the former USSR; OPEC = Organization of the Petroleum Exporting Countries.

the result is a small capital outflow, as reflected in the shift of the current account toward surplus in table 1-18. Interestingly, the capital flow reverses under the BTA policy. When the United States increases its tariffs, the reduction in trade reduces GDP in many regions (table 1-14) and leaves the U.S. economy in a relatively stronger position. The dollar strengthens in both simulations, as shown in table 1-19.

Table 1-18. Simulated Effects of U.S. Policies on Current Accounts, 2010, 2020, and 2030

Billions of dollars

Country or Group	US-Tax			US-TaxAdj		
	2010	*2020*	*2030*	*2010*	*2020*	*2030*
United States	0.8	0.2	1.8	−5.1	−4.2	0.0
Japan	−1.4	−1.0	−1.5	−1.7	−1.4	−2.6
Australia	0.3	0.4	0.5	0.8	0.9	1.1
Europe	−2.1	−1.6	−2.4	−0.3	−0.5	−2.1
Other OECD members	1.4	1.6	1.7	3.5	3.6	4.0
China	−0.6	−0.2	−0.4	−0.9	−0.2	−0.7
India	0.0	0.1	0.1	0.2	0.3	0.3
Less-developed countries	0.5	0.4	0.4	3.0	1.7	1.9
EEFSU	0.4	0.5	0.7	1.2	1.5	1.7
OPEC	0.8	0.4	0.4	0.3	−0.3	−0.4

Source: Authors' simulations.
Note: Exchange rates are measured as dollars per unit of foreign currency. OECD = Organization for Economic Cooperation and Development; EEFSU = Eastern Europe and the former USSR; OPEC = Organization of the Petroleum Exporting Countries.

Table 1-19. Simulated Effects of U.S. Policies on Real Exchange Rates, 2010, 2020, and 2030

Percentage changes from values for business as usual

Country or Group	US-Tax			US-TaxAdj		
	2010	*2020*	*2030*	*2010*	*2020*	*2030*
United States	—	—	—	—	—	—
Japan	−2.0	−2.1	−2.4	−4.6	−5.0	−5.6
Australia	−1.7	−1.8	−2.0	−3.9	−4.1	−4.4
Europe	−1.8	−1.9	−2.2	−4.2	−4.5	−5.0
Other OECD members	−2.2	−2.4	−2.6	−5.0	−5.4	−5.9
China	−1.8	−1.9	−2.2	−4.1	−4.5	−5.0
India	−1.8	−2.0	−2.3	−4.3	−4.7	−5.2
Less-developed countries	−1.8	−2.0	−2.3	−4.2	−4.7	−5.2
EEFSU	−1.8	−1.9	−2.2	−4.1	−4.4	−4.8
OPEC	−2.2	−2.5	−2.8	−4.1	−4.7	−5.2

Source: Authors' simulations.
Note: Exchange rates are measured as dollars per unit of foreign currency. OECD = Organization for Economic Cooperation and Development; EEFSU = Eastern Europe and the former USSR; OPEC = Organization of the Petroleum Exporting Countries.

As shown in table 1-20, the U.S. carbon tax causes much larger percentage changes in fuel prices than did the European tax, reflecting the lower initial energy prices in the United States. The price of coal rises by 50 to 94 percent, compared with the 23 to 33 percent increase under the European policy. Fuel consumption, in turn, falls by larger percentages; coal, for example, drops by 20 to 29 percent rather than the 8 to 13 percent in Europe. It is interesting to

Table 1-20. Simulated Effects of U.S. Policies on U.S. Prices and Output, 2010, 2020, and 2030

Percentage changes from values for business as usual

	Prices			Quantities		
Sector	*2010*	*2020*	*2030*	*2010*	*2020*	*2030*
US-Tax						
Electric utilities	6.6	7.9	9.4	−3.6	−4.3	−5.0
Gas utilities	1.1	1.2	1.4	−3.7	−4.4	−5.3
Petroleum refining	14.3	17.2	20.6	−10.9	−12.4	−13.8
Coal	59.7	75.9	94.3	−19.4	−23.7	−28.3
Crude oil	18.6	22.6	27.2	−13.2	−15.3	−19.0
Mining	0.6	0.6	0.7	−1.0	−0.8	−0.8
Agriculture	0.3	0.4	0.4	−0.3	−0.3	−0.4
Forestry and wood products	0.0	−0.1	0.0	−0.4	−0.2	−0.3
Durables	−0.1	−0.2	−0.2	−0.7	−0.4	−0.4
Nondurables	0.3	0.4	0.5	−0.2	−0.2	−0.3
Transportation	0.5	0.5	0.6	−0.4	−0.3	−0.4
Services	0.2	0.2	0.2	0.1	0.1	0.1
US-TaxAdj						
Electric utilities	6.6	8.0	9.5	−3.5	−4.2	−4.9
Gas utilities	1.1	1.1	1.3	−3.7	−4.4	−5.2
Petroleum refining	14.9	18.2	22.0	−9.1	−10.4	−11.7
Coal	59.5	75.8	94.2	−19.5	−23.9	−28.5
Crude oil	17.0	20.8	25.2	−13.3	−15.3	−18.9
Mining	0.5	0.5	0.7	−1.6	−1.4	−1.4
Agriculture	0.1	0.2	0.3	−0.6	−0.6	−0.7
Forestry and wood products	−0.3	−0.3	−0.3	−0.2	0.0	−0.1
Durables	−0.1	−0.1	0.0	−1.1	−0.7	−0.8
Nondurables	0.3	0.3	0.5	−0.3	−0.3	−0.4
Transportation	0.5	0.5	0.6	−0.3	−0.3	−0.4
Services	0.2	0.1	0.2	0.2	0.2	0.2

Source: Authors' simulations.

note that the BTAs generally do not have the mild protective effect seen under the European case. The reduction in world GDP, and the consequent drop in demand for U.S. exports, more than offsets the shift of domestic consumption from imports to domestic producers.

Conclusion

Carbon taxes on trade in primary energy commodities (that is, coal, oil, natural gas) are straightforward and would likely be part of any domestic carbon tax or permit trading system. Computing BTAs for the carbon content of all other traded goods and services, however, is very complex. In practice, it would require calculations on a country-of-origin basis for all trading partners of the country applying the BTAs. The complexities increase when a good that has been manufactured contains intermediate goods that have a number of different sources across countries. However, our results show that the tariffs would be small for most goods at moderate carbon tax levels. At an aggregate level, the adjustments for most manufactured goods would be on the order of 1 or 2 percent. However, within some narrowly defined and energy-intensive industries, such as aluminum refining, the rates would be considerably higher. Also, the adjustments are proportional to the carbon tax being imposed, so very high carbon taxes could lead to more significant BTAs.

We find that the BTAs would be effective at reducing leakage of emissions, but leakage is very small even without the BTAs. Moreover, much of the emissions gain that does occur comes about because the tariffs reduce world GDP through the overall reduction in international trade. Finally, because the BTAs are small, they have little effect on import-competing industries. We conclude that the benefits produced by BTAs for traded goods and services would be small, and they are unlikely to justify their administrative complexity or their deleterious effects on international trade.

Comments

Comment by Nils Axel Braathen

Any nonglobal policy to combat climate change would lead to some "leakage" of emissions to countries that do not participate in the "carbon coalition"—and demands for protection of the competitiveness of the most vulnerable economic sectors are to be expected. This leakage can occur through three main channels:

—via losses in competitiveness of certain sectors;

—via the markets for fossil fuels—because a reduction in fuel demand in the coalition would lead to lower world-market fuel prices and increased fuel demand in other countries;

—via changes in foreign direct investments.

The Organization for Economic Cooperation and Development (OECD) published studies of the effects of carbon taxes and border tax adjustments (BTAs) in the steel and cement sectors a few years ago—based on simulations with partial equilibrium models.[1] These simulations indicated that a nonglobal carbon tax could have a clear negative impact on the competitiveness of these sectors within the coalition—and that, in principle, BTAs could significantly reduce these effects.

For example, the steel sector study illustrated the effects of an OECD-wide tax of $25 per metric ton of carbon dioxide (CO_2). In this context, it was found that if both import taxes and export subsidies were implemented and were differentiated across steel types, and if the border tax rates were linked to emission levels in non-OECD countries, the decline in OECD steel production might be as small as 1 percent—as opposed to 9 percent if no adjustments were made. At the same time, the reduction in global emissions (5.1 percent) would be larger than without BTAs (4.6 percent). This is because the border taxes keep

---

1. For steel, see OECD (2003). For cement, see OECD (2005).

a higher share of world steel production within the OECD area, thus making more steel producers subject to the OECD-wide carbon tax.

Given their partial nature, these simulations only capture the first of the three sources of leakage listed above. Ongoing simulations with a general equilibrium model (called ENV-Linkages) at the OECD also capture the second source of leakage—and these simulations indicate that the effects via the fossil fuel markets under certain conditions are much more important than the sectoral competitiveness effects.[2,3]

However, the relative magnitude of fossil fuel market effects and sectoral competitiveness effects on total leakage seems to depend on the size of the carbon coalition. If the coalition is "large"—say, comprising all the Annex I countries—global fossil fuel demand would be reduced to a significant extent, which would trigger a significant reduction in world market fuel prices (especially for oil). This would in turn lead to a relatively strong increase in the demand for these fuels in countries outside the coalition—making this source of leakage dwarf the leakage effects stemming from a loss of sectoral competitiveness. In such a situation, BTAs would affect total carbon leakage only to a limited extent—in line with the findings of McKibbin and Wilcoxen.

If, conversely, the size of the carbon coalition is much smaller—say, only comprising the EU countries—the effects on global fossil fuel demand would be much smaller, leading to less leakage through this channel. Total leakage in this case would, however, be larger than in the previous case, due to more important sectoral competitiveness effects. In such a situation, the environmental arguments for applying BTAs could be somewhat stronger—with some important caveats, mentioned below.

While McKibbin and Wilcoxen's G-Cubed model covers only CO_2 emissions, the ENV-Linkages model also include other greenhouse gases, and the this seem to be of some importance. As long as not all relatively low-cost abatement options related to non-CO_2 greenhouse gases have been exhausted, total leakage tends to be lower when these gases are included in the analysis—as the effects are shifted from the fossil fuel markets, hence reducing the leakages that are generated through this channel, to other sectors, in particular agriculture.

It should, however, be kept in mind that the distortions created by BTAs would represent a significant economic cost—even when disregarding the very important administrative costs they would entail. As indicated by McKibbin

2. OECD (2008) provides further information about these simulations.

3. As opposed to the ENV-Linkages model, the model used by McKibbin and Wilcoxen, G-Cubed, also captures carbon leakages that stem from changes in foreign direct investment. Unfortunately, the chapter does not indicate the relative magnitude of these effects.

and Wilcoxen, to be as effective as possible, the BTAs on imports ought to be based on the carbon content of relevant products in the country where they are produced—which obviously would be very difficult to monitor and enforce in practice.

Further, the compatibility of carbon-policy-based BTAs with World Trade Organization (WTO) rules remains an open question. An overview of this issue in a recent OECD book concluded that only a WTO panel could resolve the question of the legality of such measures.[4]

Regardless of the legality of any BTA measures under the WTO, there is an important danger that the introduction of such measures could trigger tit-for-tat retaliations from the countries that would be "hit" by these measures. This could in turn have serious effects on world trade and economic development.

Hence—in line with McKibbin and Wilcoxen's conclusions—BTAs should only be considered as a last resort. The focus should instead be firmly fixed on *achieving an ambitious international approach* to address the climate change problem, with *participation by all the major greenhouse-gas-emitting countries and sectors.*

Comment by (Tom) Hu Tao

In chapter 1, McKibbin and Wilcoxen quantify the significant differences in carbon efficiency between the United States and China and India. In so doing, they provide useful analytical insights. I would like to share several comments on the issue of proposals embodied in bills like Lieberman-Warner to impose import tariffs based on the carbon content of goods.

First, when addressing the U.S. desire to impose import tariffs based on the carbon content of goods (as proposed in Lieberman-Warner-type bills) the United States' history with the Kyoto Protocol is relevant. So far, the United States is the only Annex I country that has not signed the protocol. By contrast, China and India have signed the protocol, although, because they are non–Annex I countries, they do not have legally binding obligations for greenhouse gas reductions. The post-protocol negotiation is still under way following the Bali Road Map, and there are as yet no new agreements. Given that the United States has not signed the protocol and that China and India have, it is not clear how the United States' proposed "punishment" of these countries is consistent with international agreements like the United Nations Framework Convention on Climate Change and the protocol.

4. OECD (2006, chapter 5).

Second, the Lieberman-Warner Bill and other similar bills ignore China's own domestic policies. On January 1, 2007, China implemented export tariffs ranging from 5 to 25 percent on carbon-, pollution-, and resource-intensive products, including iron, steel, coke, cement, and so on. Thus, China has already internalized the environmental costs in the prices of these products before exporting them. The tariff rate that China has applied is higher than the rate proposed in legislation. Therefore, it is unclear why the United States needs to implement a border tax adjustment (BTA) to internalize the environmental costs from China. In fact, if the United States were to implement such a bill, it would amount to the double taxation of the environmental externality for U.S. users of such products.

Third, though this proposed legislation is addressing U.S. competitiveness worries stemming from stronger environmental protection, China may also have competitiveness worries. U.S. products are not always more energy-efficient than Chinese products. More and more Chinese products have higher energy-efficiency standards than U.S. products. For example, China has adopted Euro IV vehicle emission standards, which are more stringent than both U.S. and California emission standards. Since China started the China Energy Efficiency Program two years ago, it has implemented more stringent energy standards than the United States for such products as refrigerators, air conditioners, washers, and other electric and electronic appliances. By the same logic that the United States is looking at BTAs, China could seek BTAs on the products mentioned above. Additionally, if BTAs come into vogue, China and India might view subsidized U.S. agricultural products as meriting BTAs on competitiveness grounds.

Fourth, this type of legislation is inconsistent with other aspects of U.S. policy. The Office of the United States Trade Representative (USTR), in a report submitted to Congress on China's implementation of World Trade Organization (WTO) commitments in December 2007, accused China of violating WTO rules on twelve products, including some high-carbon-content products, like coke for iron and steel. Later, the USTR warned China that it would bring a case at the WTO if China did not abolish export limits for these twelve products. The proposed legislation trying to reduce exports of high-carbon-content products from China is inconsistent with the USTR's attempts to increase exports of certain high-carbon-content products.

Comment by Arik Levinson

McKibbin and Wilcoxen's analysis in chapter 1 of this volume is prospective. It predicts future trade patterns after developed countries unilaterally impose hypothetical carbon taxes that disproportionately affect carbon-intensive industries. Their analysis is sophisticated, state of the art, and probably the best conceivable approximation of the true future effect of carbon taxes on trade, but it remains in the end a forecast—arguably not something at which economists excel.

My point, then, is that we can learn about the significance of the carbon content of trade by a *retrospective* analysis, so long as we are willing to replace as-yet-unregulated carbon for a pollutant developed countries began regulating thirty years ago. The idea is that it is easier to say what happened thirty years ago than to predict what will happen thirty years from now.

Thirty years ago, the United States unilaterally imposed strict pollution regulations that disproportionately affected pollution-intensive industries, raising fears that those industries would relocate to "pollution havens," a process now being called "leakage." Pollution-abatement operating costs for the manufacturing sector in the United States doubled as a fraction of value shipped between 1974 and 1991, but this doubling was spread unevenly across industries. For some industries (petroleum refining, primary metals, pulp and paper), costs tripled or even quadrupled. For others, pollution-abatement costs remained small or even declined. Did this change in comparative advantage across manufacturing industries lead to leakage in the past? For the evidence of that, it is useful to begin with U.S. manufacturing output for 1972 to 2001, depicted in figure 1C-1.

The top line in figure 1C-1 plots the real value shipped by U.S. manufacturers, from 1972 to 2001, indexed so the 1972 value equals 100. Real manufacturing output rose 71 percent. If over this period there were no change in the technology of abatement or production, and no change in the mix of industries making up U.S. manufacturing, then we would expect pollution emitted by U.S. manufacturers to also have risen 71 percent over this period. (Manufacturers would be producing 71 percent more of the same goods using the same methods.) But of course we know that the composition of U.S. manufacturing output has changed. America produces different goods today than it did thirty years ago—and one of the reasons might be "leakage" caused by polluting industries avoiding U.S. environmental regulations.

The bottom line in figure 1C-1 calculates the extent of the change in the composition of U.S. industries as it affects one particular pollutant, sulfur dioxide (SO_2). It uses the 1997 emissions intensities of each of the 470 industries

Figure 1C-1. The Sulfur Dioxide Content of U.S. Manufacturing, 1972–2001

Index: 1972 = 100

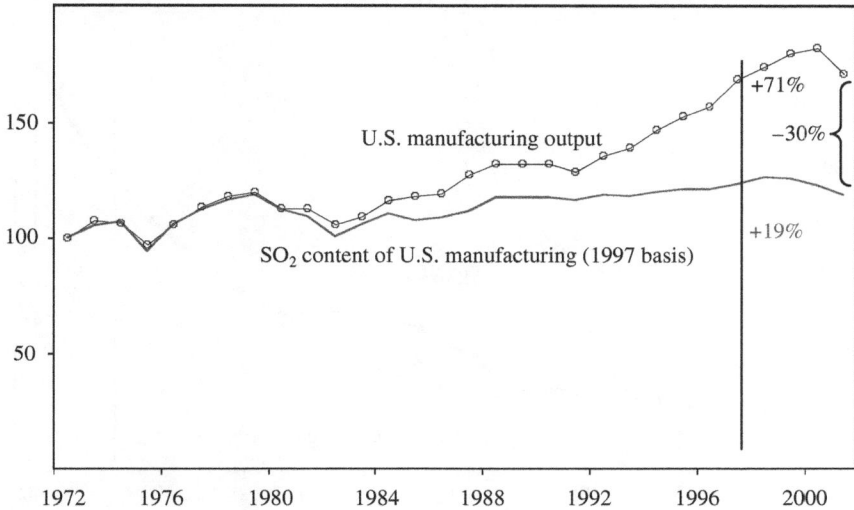

Source: Author's calculations based on NBER-CES Manufacturing Productivity Database, and EPA Trade and Environmental Assessment Model (TEAM).

that make up the manufacturing sector, as calculated by the U.S. Environmental Protection Agency. For every year, I multiplied each industry's output by its corresponding 1997 emissions intensity, and then summed the predicted SO_2 emissions across all industries. The result is the predicted amount of SO_2 that would have been emitted by U.S. manufacturing, using the 1997 technologies but the concurrent scale and mix of industries. The bottom line rises 19 percent and is lower than the 71 percent manufacturing growth for one reason: U.S. manufacturing shifted toward industries that emit less SO_2. This "green shift" of U.S. manufacturing composition resulted in SO_2 emissions that were 30 percent lower than they would have been had the mix of industries remained the same. Where did this extra SO_2 pollution go? If the SO_2-intensive industries fled to pollution havens and imported their products to the United States, we would call that leakage.

To examine whether the green shift of U.S. manufacturing might be explained by leakage, figure 1C-2 conducts exactly the same analysis but with imported manufactured goods instead of domestically produced goods. Here I am careful to account for the pollution caused by intermediate inputs to the final imports, using a Leontief-style input/output calculation similar to that used by McKibbin and Wilcoxen. The top line in figure 1C-2 depicts the real value of

Figure 1C-2. The Sulfur Dioxide Content of U.S. Imports, 1972–2001

Index: 1972 = 100

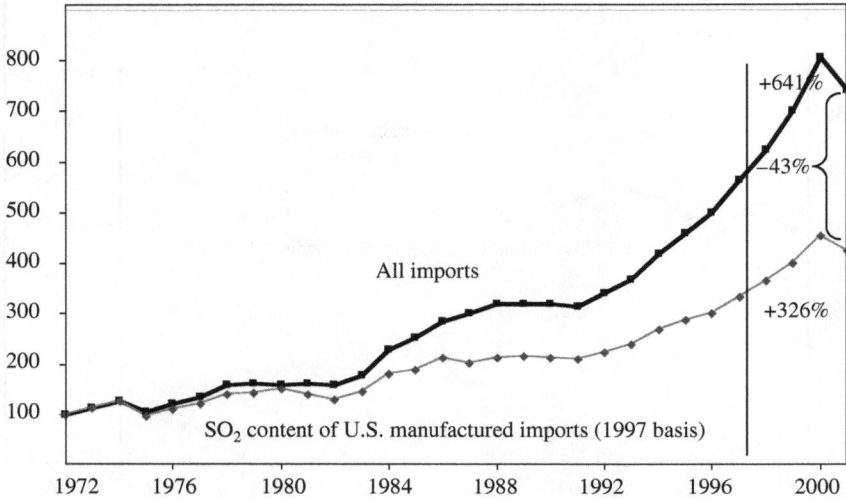

Source: Author's calculations based on NBER-CES Manufacturing Productivity Database, Center for International Data (www.internationaldata.org), and Bureau of Economic Analysis 1997 Input-Output Tables.

imports, which increased 641 percent from 1972 to 2001. The bottom line depicts the SO_2 that would have been emitted as a consequence of manufacturing those imports, had they been produced in the United States using 1997 technologies.

Figure 1C-2 depicts two noteworthy results. First, the composition of imports became cleaner over time, not dirtier. The 30 percent green shift of U.S. manufacturing was not accompanied by a corresponding "brown shift" on the part of imported goods. Instead, the composition of imports also shifted toward less-pollution-intensive goods. Second, and perhaps more startling, imports shifted toward less-polluting goods *faster* than domestic goods. The SO_2 content of imported goods was 43 percent lower that it would have been if the mix of goods being imported had remained constant.

Now, some might look at figure 1C-2 and note that U.S. imports are dominated by trade with other developed economies that were themselves enacting strict environmental regulations during this period: Canada, Japan, and the European nations. If there was leakage, perhaps the SO_2 moved to developing countries that were more likely to be pollution havens. That shift might not be

Figure 1C-3. The Sulfur Dioxide Content of U.S. Imports from Countries That Are Not Members of the Organization for Economic Cooperation and Development (OECD), 1972–2001

Index: 1972 = 100

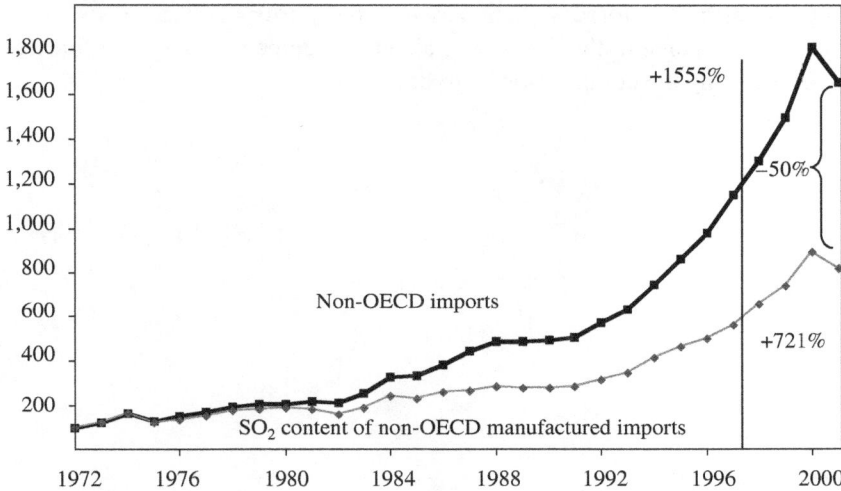

Source: Author's calculations based on NBER-CES Manufacturing Productivity Database, Center for International Data (www.internationaldata.org), and Bureau of Economic Analysis 1997 Input-Output Tables.

apparent in aggregate import data, which were composed mostly of imports from developed economies.

To address that concern, figure 1C-3 conducts exactly the same analysis but limited to imports from countries that are not members of the Organization for Economic Cooperation and Development (OECD). The figure rebuts the conjecture that leakage will be apparent in import data from less-developed countries. In fact, the green shift in imports from non-OECD countries (50 percent) was even larger than the green shift in aggregate imports (43 percent), which was itself larger than the green shift in domestic production (30 percent).

Thirty years ago, the United States began seriously regulating industrial emissions of common air pollutants such as SO_2. In the ensuing years, the U.S. manufacturing base has shifted away from production of goods that emit SO_2. But at the same time, imports to the United States, in general and from non-OECD countries in particular, have also shifted away from SO_2-intensive goods.

Note that these trends do not mean that there was no leakage of SO_2 emissions from the United States to importing countries. It could be that there was leakage, and as a consequence, the U.S. manufacturing green shift was larger than it otherwise would have been, and the imported goods' green shift was smaller. To assess that possibility, we need a general equilibrium analysis like that of McKibbin and Wilcoxen. All this analysis shows is that if there was leakage, it is not apparent in aggregate data and was swamped by other changes in the past thirty years: trade liberalization, oil prices, labor costs, and changing preferences. My forecast, then, based on this retrospective analysis, is that any carbon leakage in the future will also be swamped by as-yet-unforeseen forces affecting the composition of trade.

References

Babiker, M., and T. Rutherford. 2005. "The Economic Effects of Border Measures in Subglobal Climate Agreements." *Energy Journal* 26, no 4: 99–125.

Biermann, F., and R. Brohm. 2005. "Implementing the Kyoto Protocol Without the United States: The Strategic Role of Energy Tax Adjustments at the Border." *Climate Policy* 4, no. 3: 289–302.

Brack, D., M. Grubb, and C. Windram. 2000. *International Trade and Climate Change Policies*. London: Earthscan.

Brewer, T. 2008. "U.S. Climate Change Policy and International Trade Policy Intersections: Issues Needing Innovation for a Rapidly Expanding Agenda." Paper prepared for Seminar of Center for Business and Public Policy, Georgetown University, Washington, February 12.

Campbell J., and N. G. Mankiw. 1990. "Permanent Income, Current Income and Consumption." *Journal of Business and Economic Statistics* 8, no. 3: 265–79.

Charnovitz, S. 2003. "Trade and Climate: Potential Conflict and Synergies." In *Beyond Kyoto: Advancing the International Effort Against Climate Change*. Washington: Pew Center on Global Climate Change.

Demailly, D., and P. Quirion. 2006. "CO2 Abatement, Competitiveness and Leakage in the European Cement Industry under the EU ETS: Grandfathering versus Output-Based Allocation." *Climate Policy* 6, no. 1: 93–113.

Droge, S., and K. Kemfert. 2005. "Trade Policy to Control Climate Change: Does the Stick Beat the Carrot?" *Vierteljahrshefte zur Wirtschaftsforschung* 74, no. 2: S. 235–48.

Feldstein, M., and C. Horioka. 1980. "Domestic Savings and International Capital Flows." *The Economic Journal* 90: 314–29.

Flavin, M. A. 1981. "The Adjustment of Consumption to Changing Expectations about Future Income." *Journal of Political Economy* 89: 974–1009.

Frankel, J. 2005. "Climate and Trade: Links between the Kyoto Protocol and WTO." *Environment* 47, no. 7 (September): 8–19.

Goh, Gavin. 2004. "The World Trade Organization, Kyoto and Energy Tax Adjustments at the Border." *Journal of World Trade* 38, no. 3: 395–423.

Gordon, R. H., and A. L. Bovenberg 1996. "Why is Capital so Immobile Internationally? Possible Explanations and Implications for Capital Taxation." *American Economic Review* 86, no. 5: 1057–1075.

Hall, R. E. 1978. "Stochastic Implications of the Life-Cycle Hypothesis: Theory and Evidence." *Journal of Political Economy* 86: 971–87.

Hayashi, F. 1979. "Tobin's Marginal q and Average q: A Neoclassical Interpretation." *Econometrica* 50: 213–224.

Hayashi, F. 1982. "The Permanent Income Hypothesis: Estimation and Testing by Instrumental Variables." *Journal of Political Economy* 90, no. 4: 895–916.

Hoerner, A. 1998. "The Role of Border Tax Adjustments in Environmental Taxation: Theory and U.S. Experience." Paper presented at International Workshop on Market-Based Instruments and International Trade, Institute for Environmental Studies, Amsterdam, March 19.

Hontelez. J. 2007. "Time to Tax Carbon Dodgers" Viewpoint, BBC News (http://news.bbc.co.uk/2/hi/science/nature/6524331.stm [October 2008]).

Ismer, R., and K. Neuhoff. 2007. "Border Tax Adjustment: A Feasible Way to Support Stringent Emission Trading." *European Journal of Law and Economics* 24: 137–64.

Kopp, R., and W. Pizer. 2007. *Assessing U.S. Climate Policy Options.* Washington: Resources for the Future.

Levinson, A., and S. Taylor. 2008. "Unmasking the Pollution Haven Effect." *International Economic Review* 49, no. 1 (February): 223–54.

Lockwood, B., and J. Whalley. 2008. "Carbon Motivate Border Tax Adjustment: Old Wine in Green Bottles." NBER Working Paper 14025. Cambridge, Mass.: National Bureau of Economic Research.

Lucas, R. E. 1967. "Optimal Investment Policy and the Flexible Accelerator." *International Economic Review* 8, no. 1: 78–85.

McKibbin, W. J., and J. Sachs. 1991. *Global Linkages: Macroeconomic Interdependence and Co-operation in the World Economy.* Brookings.

McKibbin, W. J., and D. Vines. 2000. "Modelling Reality: The Need for Both Intertemporal Optimization and Stickiness in Models for Policymaking." *Oxford Review of Economic Policy* 16, no. 4: 106–37.

McKibbin, W., and P. Wilcoxen. 1997. "The Economic Implications of Greenhouse Gas Policy." In *Environment and Development in the Pacific: Problems and Policy Options,* ed. H. English and D. Runnals. Reading, Mass.: Addison-Wesley, Longman.

———. 1998. "The Theoretical and Empirical Structure of the G-Cubed Model." *Economic Modelling* 16, no. 1: 123–48.

OECD (Organization for Economic Cooperation and Development). 2003. *Environmental Policy in the Steel Industry: Using Economic Instruments* (www.oecd.org/dataoecd/58/20/33709359.pdf [November 2008]).

———. 2005. *The Competitiveness Impact of CO_2 Emissions Reduction in the Cement Sector* (www.olis.oecd.org/olis/2004doc.nsf/LinkTo/nt0000a252/$file/jt00194233.pdf [November 2008]).

———. 2006. *The Political Economy of Environmentally Related Taxes* (www.oecd.org/env/taxes/politicaleconomy [November 2008]).

———. 2008. *Climate Change Mitigation: What do we do?* (www.oecd.org/dataoecd/31/55/41751042.pdf [November 2008]).

Pearce, D., and W. McKibbin. 2007. "Two Issues in Carbon Pricing: Timing and Competitiveness." Lowy Institute Working Paper in International Economics 1.07 (www.lowyinstitute.org/Publication.asp?pid=575 [October 2008]).

Stiglitz, J. 2006. "A New Agenda for Global Warming" *Economists Voice,* July (www.bepress.com/ev [October 2008]).

Treadway, A. 1969. "On Rational Entrepreneurial Behavior and the Demand for Investment." *Review of Economic Studies* 3, no. 2: 227–39.

Uzawa, H. 1969. "Time Preference and the Penrose Effect in a Two Class Model of Economic Growth." *Journal of Political Economy* 77: 628–52.

JASON E. BORDOFF 2

International Trade Law and the Economics of Climate Policy: Evaluating the Legality and Effectiveness of Proposals to Address Competitiveness and Leakage Concerns

There is a growing consensus that a market mechanism that puts a price on carbon, such as a cap-and-trade system or a carbon tax, should be at the heart of the most flexible and cost-effective way to address climate change.[1] Ideally, such an approach would be adopted as part of a multilateral agreement. The reason is that carbon is a global pollutant, so a ton of carbon emitted in Beijing contributes to climate change just as much as a ton of carbon emitted in New York. This tragedy-of-the-commons nature of climate change raises concerns that any unilateral effort by the United States to put a price on carbon could disadvantage U.S. industrial firms or undermine the measure's environmental objective. These two concerns, in effect flip sides of the same coin, are referred to as "competitiveness" and "leakage," respectively. The competitiveness concern is that U.S. products—particularly carbon-intensive ones like steel, cement, chemicals, glass, and paper—will be at a competitive disadvantage relative to foreign-made goods if the United States unilaterally imposes a carbon price policy and thus raises production costs for U.S. firms.[2]

For helpful comments and discussion, the author would like to thank Joseph Aldy, Joel Beauvais, Steven Charnovitz, Manasi Deshpande, Elliot Diringer, Douglas Elmendorf, Andrew Guzman, Michael Levi, Bryan Mignone, Robert Novick, Warren Payne, Billy Pizer, Andrew Shoyer, Timothy Taylor, and Mark Wu. He especially thanks Pascal Noel for exceptionally valuable research assistance. Leandra English and Julie Anderson also provided helpful editorial assistance.

1. For a detailed discussion about why a market mechanism is preferable to alternative approaches, see Furman and others (2007).

2. A recent study by the Peterson Institute identifies the following carbon-intensive manufacturing industries that compete with foreign producers: ferrous metals (iron and steel), nonferrous metals (aluminum and copper), nonmetal mineral products (cement and glass), paper and pulp, and basic chemicals. See Houser and others (2008). The Lieberman-Warner Bill specifically names "iron, steel, aluminum, cement, bulk glass, or paper" as "primary products," though it permits the administrator to include "any other manufactured product that is sold in bulk for purposes of fur-

The second, related concern—emissions leakage—occurs when a policy that raises the price of carbon-intensive domestic goods causes domestic production to shift abroad and domestic consumption to shift to more carbon-intensive imports, thus undermining the policy's effect on reducing global levels of greenhouse gases (GHGs). Leakage also may occur as a result of reduced domestic demand for fossil fuel products, which depresses fuel prices in the global market and thus results in increased consumption.

An often-proposed response to the related concerns about competitiveness and leakage, which indeed has been incorporated into the leading cap-and-trade legislation, is to level the carbon playing field and encourage developing countries to adopt climate change policies by imposing a border adjustment that puts a price on the carbon contained in imports from countries without similarly stringent climate policies. Under a cap-and-trade system, this border measure could take the form of a requirement that importers from countries without comparable emissions reduction policies purchase emissions allowances to cover the carbon content of their products (or, alternatively, pay a tax equal to the allowance price). In theory, U.S. exporters might also be provided with allowances as rebates for the price of the embedded carbon in their products (though no proposal today calls for this).

Though perhaps sound in theory, the wisdom of leveling the carbon playing field by imposing border adjustments is more debatable when the expected benefits are weighed against the potential harms. The second section of this chapter briefly outlines these benefits and harms, and finds that one of the oft-cited benefits (and one that is most relevant under international law)—the reduction in GHG emissions—is likely to be quite small. To help fully explain the expected costs, and thus better compare them with the expected benefits of competitiveness and leakage prevention measures, the third section then analyzes one particular concern regarding the compatibility with World Trade Organization's (WTO's) law of border adjustments. Given space constraints, this chapter does not explore all the novel issues or claims that might be raised in evaluating such a complex legal question. Rather, the purpose of this section is to highlight the key questions that a WTO panel would raise in its analysis, focusing on how that legal analysis should be informed by the economics of a cap-and-trade system. As with any complex legal question, it is difficult to predict with any certainty how a WTO panel would rule, but the

ther manufacture" and generates significant greenhouse gas (GHG) emissions during production. America's Climate Security Act of 2007, S 2191, 110th Cong. [hereafter ACSA], Sec. 6001 (10). Unless otherwise noted, references to Lieberman-Warner throughout do not reflect revisions in the manager's substitute amendment, released May 21, 2008 (http://epw.senate.gov/public/index.cfm?FuseAction=Files.View&FileStore_id=aaf57ba9-ee98-4204-882a-1de307ecdb4d [October 2008]).

section identifies several ways in which a border adjustment might not be compliant with WTO law. Viewing border adjustments through the lens of WTO law also raises broader questions about the wisdom of imposing border adjustments as a policy matter. As an alternative, some have proposed the use of free allocation to address competitiveness concerns and political exigencies, and the fourth section considers the desirability and WTO compatibility of that approach, concluding that although free allocation, depending on its design, may be WTO compliant, that is precisely because it will be largely ineffective in protecting U.S. industries and workers, instead effectively constituting a transfer from government to shareholders of firms. The fifth section concludes that the expected costs from both border adjustments and free allocation may well outweigh the benefits, and it suggests alternative mechanisms to address climate change while mitigating leakage and adverse effects on workers in carbon-intensive sectors.

The Expected Benefits and Harms of Border Adjustments

Weighing the expected benefits of border adjustments against the expected harms raises doubts about the wisdom and effectiveness of such measures. As to the expected benefits, there are at least three. First, the environmental benefit of border adjustments would be to avoid some of the increase in foreign emissions that would otherwise occur in response to a unilateral U.S. climate policy. This potential increase in foreign emissions (that is, leakage) is small, however. Though estimates vary, most suggest that roughly 10 percent of the reduction in U.S. emissions will be replaced by increases in foreign emissions.[3] Most U.S. emissions occur in nontradable sectors, such as transport and residential housing. Further, some firms use little energy relative to other factors that may be more important in determining the location of trade.[4] Even in carbon-intensive sectors, it is estimated that production will decline in response

3. The Environmental Protection Agency estimates U.S. emissions leakage rates under Lieberman-Warner of approximately 11 percent in 2030 and 8 percent in 2050. "EPA Analysis of the Lieberman-Warner Climate Security Act of 2008 S 2191 in 110th Congress, March 14, 2008" [hereafter EPA Analysis S 2191], 84 (www.epa.gov/climatechange/downloads/s2191_EPA_Analysis.pdf [February 2009]). Paltsev (2001) estimates leakage rates of 10.5 percent from Annex I countries under their Kyoto caps, though he estimates U.S. leakage rates (under never-ratified Kyoto targets) of only 5.5 percent. McKibbin and others (1999) estimated in 1999 that if the U.S. unilaterally adopted Kyoto targets, leakage rates would be roughly 10 percent in 2010. The Intergovernmental Panel on Climate Change (2001) surveys a number of multiregional leakage estimates, finding a range of 5 to 20 percent.

4. For example, energy costs in most manufacturing industries are less than 2 percent of total costs; Morgenstern and others (2007).

to a carbon price more because of a reduction in domestic consumption than because of a shift to imports or the offshoring of production.[5]

More important, according to a recent analysis by the U.S. Environmental Protection Agency, a border adjustment on carbon-intensive manufactured imports, like that proposed in the Lieberman-Warner Bill, would reduce that 10 percent by only about half a percentage point because it (1) ignores production leakage due to export competitiveness, (2) applies only to a subset of imports, and (3) does not address the increased global demand for fossil fuels in response to the lower prices that reductions in the U.S. quantity demanded will have.[6] The economists Warwick McKibbin and Peter Wilcoxen similarly find that border adjustments "would reduce leakage of emissions reductions very modestly" (see chapter 1 in this volume).

To keep the environmental benefit of preventing leakage from carbon-intensive industries in perspective, consider that only 6 percent of total U.S. emissions comes from these industries.[7] Moreover, if the United States unilaterally implements a border adjustment, it is easy to envision other countries reshuffling their trade to avoid the border charge. For example, the United States might import more from Europe and less from Brazil, China, and India, while these developing countries just send more to Europe.

Even if border adjustments do little to reduce emissions leakage, some argue that they can enhance the overall environmental utility of a cap-and-trade regime by achieving greater global GHG reductions in two ways. One argument is that border adjustments will induce developing countries to reduce their own emissions so as to avoid the border charge. Given that China alone is expected to account for 47 percent of the growth in GHG emissions over the

5. Aldy and Pizer (2008).

6. EPA analysis S 2191, 84. In a scenario where Annex II countries take no action on their own, but the United States unilaterally adopts an emissions reduction policy, the International Reserve Allowance Requirement in the Lieberman-Warner Climate Security Act reduces leakage from 361 metric tons of carbon dioxide equivalent (MtCO2e) to 350 MtCO2e in 2030 (or from 11.6 percent of U.S. reductions to 11.3 percent) and from 412 MtCO2e to 385 MtCO2e in 2050 (or from 8.2 percent of U.S. reductions to 7.6 percent). The EPA's ADAGE model does not allow it to break out how much of the emissions leakage is from each of these various sources. In his paper measuring the emissions leakage from implementing the Kyoto Protocol, however, Paltsev (2001, 68 n. 4) finds that leakage from Annex I demand reductions, which lead to reduced world prices and thus increased Annex II consumption, accounts for about one-quarter of total leakage. It is important to note that the manager's substitute amendment to Lieberman-Warner expands the definition of "covered products" to include not only primary carbon-intensive goods but also manufactured goods for consumption that generate a substantial quantity of direct and indirect GHG emissions. Sec. 1311 (7) and (14). Even if such broader coverage did more to reduce leakage, it could create enormous administrative challenges. For most downstream goods, however, a carbon price is likely to be a small enough component of total cost that a border adjustment would do little to change trade flows.

7. Houser and others (2008, xiv).

next twenty-five years,[8] engaging emerging economies in efforts to address climate change is critically important. Yet only a very small fraction of carbon-intensive products made in China are exported to the United States, so a border adjustment in the United States would be a small stick with which to pressure China to implement more costly low-carbon production processes. Though China accounts for one-third of global steel production, less than 1 percent is sold to the United States; the U.S. market also accounts for just 3 percent of Chinese aluminum production, 2 percent of paper production, and less than 1 percent of both basic chemicals and cement.[9] Moreover, even if border adjustments were successful in reducing emissions in carbon-intensive industries, they would do nothing to reduce the three-quarters of Chinese emissions that come from other sources.[10]

The other argument that border adjustments can enhance the environmental effectiveness of a cap-and-trade program concerns the manner in which foreign firms are permitted to acquire emissions allowances for the importation of carbon-intensive goods. Lieberman-Warner permits firms to buy allowances at a set price from a separate pool outside the domestic cap, which in effect would be a carbon tax. But it also permits a firm to submit allowances from other comparably effective cap-and-trade systems, such as that in the European Union. These allowances would then be retired, thereby reducing total GHG emissions by an equivalent number of tons. There are three reasons, however, such a provision would, in fact, do little to increase the environmental effectiveness of the U.S. cap-and-trade system. First, it only applies if the price of U.S. permits exceeds the price of permits in other cap-and-trade systems, because otherwise a foreign firm will find it cheaper to buy allowances from the U.S. pool. Second, if a Chinese firm buys an allowance from the EU system for 1 ton of GHG and retires it by submitting it to the United States to satisfy its import requirement, that would have the effect of tightening the EU cap by 1 ton. In effect, U.S. law would thus be expanding the scope of the EU's cap-and-trade regime to cover carbon-intensive goods in non-EU countries. That is going to drive up the price of permits in the EU and force EU firms and consumers to satisfy a more stringent cap than they imposed on themselves. The likely response from the EU would then be to relax its cap so that EU countries themselves were still obligated to meet only the emissions target they originally set for themselves, which would negate any greater emissions reductions resulting from the U.S. border adjustment provision. Third and finally, it

8. Author's calculations, based on DOE (2008).
9. Houser and others (2008, xvi).
10. This estimate is based on a personal communication with Trevor Houser, Peterson Institute for International Economics, July 14, 2008.

is not clear that the legislation creating the EU's cap-and-trade system permits foreign entities to buy allowances in the EU system unless the EU affirmatively allows it as part of an agreement to link emissions trading schemes.[11]

The second potential benefit of a border adjustment is that it can protect certain industries by leveling the carbon playing field relative to carbon-intensive imports. The Environmental Protection Agency, for example, estimates that U.S. imports from Annex II countries—those not subject to the Kyoto Protocol caps—would be roughly 12 percent higher in 2050 without a border adjustment than they would be with one.[12] To some extent, the benefits to U.S. carbon-intensive manufacturers may be limited by the fact that many of the carbon-intensive imports to the United States come from Annex I countries— those that (with the exception of the United States) are already part of the Kyoto Protocol and thus would likely be exempt from most border adjustment proposals.[13] Indeed, Canada is the largest source of imports in all carbon-intensive industries except one, with Europe and Russia not far behind.[14] At the same time, however, the competitiveness benefit may still be considerable because the sectors in which roughly two-thirds of U.S. imports come from Annex II countries (chemicals and cement) are also among the carbon-intensive sectors that comprise the largest shares of U.S. gross domestic product (GDP) and employment.[15] Moreover, the growth rates for imports in these sectors have been more rapid than for imports in other carbon-intensive sectors.[16]

Third, as a political matter, border adjustments also may have the benefit of helping to secure passage of a cap-and-trade bill in the U.S. Congress (where some measure to address adverse effects on domestic industry likely will be necessary). They might also encourage other developed nations to adopt sim-

11. European Commission (EC), "Directive 2003/87/EC of the European Parliament and of the Council of 13 October 2003 Establishing a Scheme for Greenhouse Gas Emission Allowance Trading within the Community and Amending Council Directive 96/61/EC, Art 12 & 25," *Official Journal of the European Union*, L 275/32, 2003.

12. EPA Analysis S 2191, 85. See also Morgenstern and others (2007).

13. Annex I countries account for 54 percent of U.S. steel imports, 78 percent of aluminum imports, 34 percent of chemicals imports, 87 percent of paper imports, and 35 percent of cement imports. See Houser and others (2008, 44). To be sure, some Annex I countries like Canada may fail to meet their targets and thus may not be judged to have taken comparably effective measures, even though they are subject to the Kyoto Protocol's caps.

14. Houser and others (2008, 44). Note that to the extent products from these countries already internalize a carbon price, U.S. products may be viewed as receiving a subsidy by emitting carbon without paying such costs. See Stiglitz (2006a, 2006b).

15. Houser and others (2008, 11). Chemicals and cement make up 1.68 and 0.43 percent, respectively, of U.S. GDP and 0.65 and 0.38 percent, respectively, of employment. Paper has roughly equal shares to cement: 0.44 percent of GDP and 0.36 percent of employment. Steel and aluminum make up only 0.29 and 0.20 percent, respectively, of GDP and 0.19 and 0.11 percent, respectively, of employment. Houser and others (2008, table 1.2).

16. Houser and others (2008, 46, figure 3.3).

ilar policies, which might do more to induce developing countries to negotiate an international agreement.

Against these expected benefits need to be weighed at least three expected costs of border adjustments. First, there is a risk that the border adjustment system could be abused for purely protectionist reasons by U.S. firms facing growing global competitive pressures. Second, there is a risk that border adjustments could lead to retaliatory tit-for-tat trade wars, particularly with developing nations, which may believe that developed nations bear a greater responsibility for curbing climate change. India or China, for example, could well argue that the United States bears a greater responsibility for cumulative emissions and is still a much larger GHG emitter on a per capita basis. Moreover, to this point, the United States has taken relatively little action to address climate change compared with many Kyoto countries, and there is a risk that any eventual climate change policy would have limited effectiveness once Americans understand the true impact of cap-and-trade on energy prices and political pressure then builds to ease that pain. In that case, introducing border adjustments as a legitimate tool to address climate change may encourage other nations that are doing more to curb emissions, such as those in the European Union, to impose such border adjustments on the United States. Border adjustments for a carbon price could also set a dangerous precedent for the use of border tax adjustments to compensate for other competitive disadvantages seemingly imposed on domestic producers, such as minimum wage or health care regulations. Such risks to free trade, which delivers $1 trillion in benefits annually to the U.S. economy,[17] are particularly harmful at a time when America's commitment to free trade is ever more in doubt.[18] Finally, there is a risk that a border adjustment would be illegal under WTO law, as discussed in the next section, which could potentially lead the WTO to authorize retaliatory tariffs.

Evaluating Border Adjustments under WTO Law

The WTO is the international organization responsible for overseeing the multilateral trading system. It was created in the Uruguay Round of multilateral trade negotiations out of what had previously been the General Agreement on Tariffs and Trade (GATT) institutional structure. The WTO also consists of a treaty that combines a variety of detailed agreements, including the GATT.

17. Bradford, Grieco, and Hufbauer (2005).

18. According to a recent Pew Research Center poll, a 48 percent plurality said that free trade agreements are a bad thing for the country, compared with 35 percent of the public, who call them a good thing. In July 2004, the positions were reversed, with 47 percent of respondents calling free trade agreements positive and 34 percent calling them negative. See Pew Research Center (2008).

The WTO has a dispute settlement system, under which an allegation of a violation of one or more of these agreements can be brought before a WTO panel and, on appeal, to the WTO Appellate Body. If the losing nation fails to adhere to the WTO's ruling, the complaining nation may seek authority to impose retaliatory tariffs. Because such a remedy, which precludes retaliation if the offending provision is cured, lacks deterrent power, "many governments engage in trade or economic policies that test the limits of WTO law. This pattern of behavior ought to be kept in mind in considering the extent to which WTO rules lacking clarity should constrain the design of climate policies."[19]

There are three steps in the analysis of whether a border adjustment is consistent with WTO law. First, is the border adjustment consistent with WTO market access commitments? If so, is it also consistent with the nondiscrimination obligations under the WTO? If not, is it permissible nonetheless under one of the exceptions provided for under GATT Article XX?[20]

As to the first question, GATT Article II prohibits tariffs above a particular ceiling, and Article XI generally prohibits quantitative restrictions on imports. A border adjustment that applies to imports the same requirements imposed on domestic products is generally permissible as a border-enforced internal measure, assuming it does not violate national treatment or most-favored-nation treatment obligations (discussed below).[21] Assuming the border adjustment is imposed as part of an overall domestic cap-and-trade system, therefore, the WTO may well view it as a border-enforced internal measure.[22] The focus of this analysis, therefore, is on the second and third questions regarding whether the measure is discriminatory or falls under an environmental exception.

The analysis assumes that the United States adopts a cap-and-trade system and requires importers to purchase emission allowances at the U.S. market price, which seems the likeliest form of border adjustment given current policy discussions. Much of the analysis of the GATT's nondiscrimination provisions and environmental exceptions would be the same even if the United States

19. Charnovitz (2003, 144).

20. Other exceptions also exist, such as GATT Article XXI's security exceptions, though only Article XX is likely to be relevant for the purposes of border adjustments.

21. GATT Article III explains: "Any internal tax or other internal charge, or any law, regulation or requirement of the kind referred to in paragraph I which applies to an imported product and to the like domestic product and is collected or enforced in the case of the imported product at the time or point of importation, is nevertheless to be regarded as an internal tax or other internal charge, or a law, regulation or requirement of the kind referred to in paragraph 1, and is accordingly subject to the provisions of Article III."

22. As discussed further below, however, a border adjustment may fall outside the scope of GATT Article III, because border-enforced internal measures can be applied to like "products," but a requirement to hold emission allowances may be considered a charge not on the "product," but rather on the process or production method (PPM).

adopted a carbon tax and imposed a carbon tax on imports. Where important differences in the legal analysis exist, however, those will be noted.

Nondiscrimination Obligations

Even if a border adjustment is accepted as a permissible border-enforcement of an internal measure, the border adjustment must also not violate Article III's "national treatment" obligation by discriminating against imports or Article I's "most-favored-nation" obligation by discriminating among importing nations. These two requirements are discussed in turn.

NATIONAL TREATMENT. Article III:4 requires that the United States accord to imported products "treatment *no less favorable* than that accorded to *like* products of national origin in respect of all laws, regulations and requirements affecting their internal sale, offering for sale, purchase, transportation, distribution or use" (emphasis added). This subsection first discusses the meaning of "like products" and then whether, even if a border adjustment were found not to discriminate against "like products," the amount of the border adjustment could be determined in a nondiscriminatory fashion. It concludes by briefly noting how the foregoing analysis might differ if a domestic cap-and-trade regime were viewed not as an internal regulation, covered by Article III:4, but rather an internal tax, covered by Article III:2.

The principle behind Article III is straightforward: A member cannot treat imported goods worse than domestic goods. In the case of climate change border adjustments, however, this seemingly straightforward principle proves exceptionally difficult to put into effect because the same goods from a global trade standpoint may be very different from a climate change standpoint if one is much more carbon-intensive than the other.

The Appellate Body has explained that whether two products are "like" under Article III:4 is to be determined by whether they are in a "competitive relationship,"[23] and thus a basic industrial product like steel would most likely be considered "like" other steel, even if they were produced in ways that emitted different amounts of carbon. An importer of more carbon-intensive steel might thus challenge a border adjustment that required it to purchase more allowances to reflect the higher carbon content by claiming its "like" product was being treated less favorably. If the U.S. regulation instead imposed an allowance requirement equal to that paid by U.S. manufacturers regardless of carbon content (for example, a charge per unit of steel imported), low-carbon producers (such as those in nations that rely more heavily on nuclear or nat-

---

23. WTO, *Appellate Body Report on European Communities: Measures Affecting Asbestos and Asbestos-Containing Products*, WT/DS135/AB/R, March 12, 2001 [hereafter *EC-Asbestos*], paragraph 99.

ural gas) would likely object on the grounds that their products were being treated less favorably.

The United States might respond that high-carbon steel is not "like" lower-carbon steel because one contributes more than the other to climate change. Generally speaking, the interpretation of "like" products does not permit differentiation based on the way a product is made (so-called process and production methods, or PPMs), but rather only on the product's physical characteristics.[24] Thus, the Appellate Body found that chrysotile asbestos fibers were not "like" fibers made from other materials given the public health risks of asbestos.[25] By contrast, tuna caught in a dolphin-friendly way was "like" tuna caught in a dolphin-unfriendly way.[26] Given that steel created in a climate-friendly way is physically indistinguishable from steel created in a climate-unfriendly way, GATT jurisprudence suggests that a measure that distinguishes like products based on how much carbon was emitted in their creation might not fall within the scope of Article III.[27]

The distinction in GATT jurisprudence between a product, on the one hand, and PPM, on the other hand, need not be fatal to a carbon border adjustment's legality, however. WTO case law suggests that PPM distinctions between like products are most likely not permissible under Article III but may be permitted under Article XX.[28] Indeed, in the *US-Gasoline* and *US-Shrimp* cases, the Appellate Body ruled that PPM restrictions were not necessarily inconsistent with the GATT because they fell within the scope of Article XX.[29]

Even if a border adjustment were found not to discriminate between like products, importers would need to pay the same price per ton of carbon emitted as domestic producers (through the purchase of allowances in a market) to be treated "no less favorably." The problem, however, is that it can be difficult to agree on what price U.S. manufacturers paid to emit a ton of carbon under a domestic cap-and-trade scheme.

24. Matsushita, Schoenbaum, and Mavroidis (2003, 163); Hudec (2000, 187, 191).

25. *EC-Asbestos* AB.

26. GATT, *United States: Restrictions on Imports of Tuna*, GATT BISD (39th Supp.), at 155 (1993), reprinted in 30 ILM 1594 (1991) (unadopted); GATT, *United States: Restrictions on Imports of Tuna*, DS29/R, June 16, 1994, reprinted in 33 ILM 839 (1994) (unadopted).

27. Bhagwati and Mavroidis (2007).

28. See Charnovitz (2002); Hudec (2000, 192).

29. WTO, *Appellate Body Report on United States: Standards for Reformulated and Conventional Gasoline*, WT/DS2/AB/R, May 20, 1996 [hereafter *US-Gasoline AB*], 13–22; WTO, *Appellate Body Report on United States: Import Prohibition of Certain Shrimp and Shrimp Products*, WT/DS58/AB/R, May 15, 1998 [hereafter *US-Shrimp AB*], paragraphs 141–49; WTO, *Appellate Body Report on United States: Import Prohibition of Certain Shrimp and Shrimp Products* (Implementation under Article 21.5), WT/DS58/RW/AB, November 21, 2001 [hereafter *US-Shrimp Article 21.5 AB*], paragraphs 149–52.

There are several ways in which the economic incidence of allowance costs may be in dispute, but just consider the widely discussed question of whether allowances should be auctioned or freely allocated.[30] It is often assumed that to the extent allowances are freely distributed, the price charged to importers would need to be discounted proportionally.[31] The problem with this approach, however, is that regulated entities, say upstream importers or extractors of fossil fuels, will still pass on allowance costs to firms and consumers even if they receive allowances for free.[32] The reason is that allowances, even ones received for free, can be sold for cash in a liquid secondary market and thus there is an opportunity cost to using one to emit a ton of carbon. A GHG emitter will decide not to sell an allowance only if it can recoup that opportunity cost, which happens by raising prices. In such a scenario, carbon-intensive manufacturers (and their customers) would still bear the full market price for emitting carbon, and thus there would be no reason to reduce that price for importers. Indeed, even if the manufacturers themselves received the free allowances, they would still pass the opportunity cost of using allowances to emit carbon on to their customers. As discussed in greater detail below, that is precisely why the Congressional Budget Office and other analysts think that free allowances have essentially the same effect on emissions and output as auctioned allowances. Thus, it would not disadvantage importers to pay the market price for carbon even if domestic manufacturers received free allowances themselves. That is why allowance allocation is a distributional issue that should be separated from the issue of compliance obligations under the cap.[33]

Finally, even if the right price could be determined, that carbon price would need to be imposed as a border adjustment *based on the carbon content of the import*, which can be exceptionally complicated to determine. Foreign manufacturers asked to provide detailed carbon content information may be unwilling to do so, or even unable given increasingly disaggregated global supply chains

30. Another complication regarding the economic incidence of allowance costs arises in cost-of-service regulated electricity markets. In such markets, the full price of carbon may not be passed on to manufacturers, and thus it would discriminate against foreign imports to charge them the market price for emissions allowances.

31. See, for example, Pauwelyn (2007, 22).

32. See Congressional Budget Office (2003, 2007b). Notably, the incidence of a border adjustment would parallel the incidence of a domestic cap-and-trade program. In a domestic cap-and-trade regime, the statutory incidence would fall on the firms required to surrender allowances at the end of the year, while the economic incidence falls primarily on downstream consumers of energy and final products (because the demand for energy is so inelastic). With a border adjustment, the foreign firm exporting goods to the United States would face the statutory incidence—it would actually have to buy the allowances or pay the government-imposed border charge—while the economic incidence falls mostly on American consumers, who will see the cost of the border adjustments priced into the final retail price of the goods they buy.

33. Kopp (2007).

for production.[34] In that case, the United States might calculate the border adjustment based on external industry-wide benchmarks, as proposed for the U.S. British thermal unit tax in 1993. In *US-Gasoline*, however, the GATT Panel struck down a U.S. regulation assigning foreign producers a standard baseline while domestic refiners got an individual one.[35] In the case of industrial goods, in which the amount of carbon emitted can vary dramatically (depending on such factors as the source of energy, such as nuclear versus coal, and the production process, such as lower-carbon steel mini-mills versus higher-carbon integrated mills), applying one baseline carbon content to every product regardless of how and where it was produced may well be considered discriminatory.

Although the above discussion has largely discussed a cap-and-trade system as a domestic regulation, the WTO might alternatively view it as falling under the national treatment requirements of Article III:2, which concerns "internal taxes or other internal charges of any kind" that are applied "directly or indirectly" on products. The requirement to purchase allowances that force firms to internalize the social cost of the carbon they emit may be viewed as effectively the same as a carbon tax. Indeed, from an economic perspective, if there were complete certainty about the costs and benefits of a carbon price, there is little difference between a carbon tax and a cap-and-trade system. If the government were to issue the precise number of permits so that the market settled on a value of $15 to emit a ton of carbon, that would be the same as setting a $15 per ton carbon tax. In reality, however, there is considerable uncertainty about the costs of climate change and of policies to mitigate it. Quantity instruments like cap-and-trade systems provide certainty about how much emissions will be reduced but uncertainty about the costs, whereas price instruments like carbon taxes provide certainty about costs but uncertainty about just how much emissions will be reduced.

Viewed in this way, the carbon price signal created by requiring the remission of an allowance to emit a ton of carbon might be viewed as equivalent to a tax,[36] and the requirement for importers to buy an allowance for the carbon content of their products may be judged "a charge equivalent to an internal tax."[37] In that case, many of the same questions would exist as to whether a

34. See Krugman (2008). As difficult as it is to determine the carbon content of carbon-intensive manufactured goods, it is vastly more complicated to do so for manufactured goods for consumption made from those carbon-intensive primary goods, as required in the Manager's substitute amendment to Lieberman-Warner, Sec. 1311 (7) and (14).

35. The Appellate Body rejected the U.S. defense that data from foreign gasoline producers was unverifiable, though it agreed with the panel's suggestion that using a statutory baseline might be permissible when "the source of imported gasoline could not be determined or a baseline could not be established because of an absence of data." *U.S.-Gasoline AB*, 27.

36. See Ismer and Neuhoff (2004, 4–8).

37. GATT Article II:2(a).

tax on the process or production method (so-called hidden taxes, or *taxes occultes*), rather than on inputs incorporated into the final product, may be adjusted at the border.[38] Notably, a GATT panel in the *US-Superfund* case permitted the United States to impose a domestic tax on certain chemicals on imports that had used the same chemicals "as materials in the manufacture or production" of these imports, though the panel did not address whether these chemicals had to be physically present in the imported product.[39] The specific question whether hidden or process taxes may be adjusted at the border, however, was left unanswered by a 1970 GATT working group on the issue.[40] Even if a carbon tax is judged to be adjustable at the border, it would still have to meet the national treatment obligations of Article III, and many of the concerns discussed above would still exist.

MOST-FAVORED-NATION TREATMENT. The second nondiscrimination obligation a border adjustment must satisfy is Article I's "most-favored-nation" requirement, which prohibits discrimination between WTO members. Border adjustment proposals typically only apply to imports from countries that do not have a comparably effective climate policy already in place, because otherwise imports would effectively be paying a carbon price twice. Yet such an approach would seem to violate Article I because it would be treating two "like" products differently depending on their origin. The United States might argue that the treatment is nondiscriminatory because the restriction is based not on origin but on conditions of production that apply equally to all nations, and that the treatment differs only because the objective of mitigating climate change is being met differently in different places.[41] Even supporters of border adjustments, however, recognize that such a claim would face difficulty.[42] Indeed,

38. For a discussion of these issues, see Pauwelyn (2007, 19–20).

39. GATT, *Panel Report on United States: Taxes on Petroleum and Certain Imported Substances*, GATT BISD 34S/136, June 17, 1987, paragraphs 2.5 and 5.2.4.

40. "It was generally felt that while this area of taxation was unclear, its importance—as indicated by the scarcity of complaints reported in connection with adjustment of taxes occultes—was not such as to justify further examination." GATT, GATT *Working Party Report on Border Tax Adjustments*, GATT BISD 18S/97, December 2, 1970, paragraph 15. Pauwelyn (2007, 20–21) provides an argument that a carbon tax should be adjustable at the border.

41. In the Canada-Automobiles decision, for example, the panel suggested that origin-neutral criteria might be permissible under Article I. WTO, *Appellate Body Report on Canada: Certain Measures Affecting the Automotive Industry*, WT/DS139/AB/R, WT/DS142/AB/R, February 11, 2000 [hereafter *Canada-Automobiles*], paragraphs 3.22–3.24. But see WTO, *Panel Report on Indonesia: Certain Measures Affecting the Automobile Industry*, WT/DS54, 59 and 64/R, July 23, 1998, paragraph 14.143, ruling that "GATT case law is clear to the effect that any . . . advantage [here tax and customs benefits] cannot be made conditional on any criteria that is not related to the imported product itself."

42. Memorandum from Andrew W. Shoyer, "WTO Background Analysis of International Provisions of U.S. Climate Change Legislation," February 28, 2008 (http://energycommerce.house.gov/cmte_mtgs/110-eaq-hrg.030508.Morris-testimony.pdf [February 2009]).

Article I covers not only de jure but also de facto discrimination, which the Appellate Body found to exist in the *Canada-Automobiles* case, even though the challenged measure was facially origin-neutral.[43] Because "the MFN [most-favored-nation] obligation under the GATT is unconditional and quite broad,"[44] there is good reason to believe the WTO would find a violation if border adjustments applied only to certain countries.

Moreover, even if the WTO permitted differential treatment, it would be very difficult to determine which countries have comparably effective climate policies in a way that did not give rise to discrimination claims. The European Union's cap-and-trade system, for example, covers only half the economy. Many EU member countries that impose carbon taxes have exempted energy-intensive industries.[45] Moreover, other nations (like Japan) might eschew market mechanisms altogether in favor of command-and-control regulations. It is also possible to envision ways in which governments could modify their tax systems—effectively doing a corporate tax swap—that would have little or no effect on emissions but would satisfy an assessment of comparable climate policy burdens. For example, a government could cut excise taxes on fossil fuels while imposing a carbon tax. The after-tax cost of using fossil fuels by, say, steel firms would be unchanged, but the country could argue that it has implemented a climate policy comparable to that of the United States. In theory, rather than divide countries into two groups—those with and without comparably effective climate policies—all importers might be required to pay the difference between the U.S. market price for allowances and whatever carbon price they paid in their home country. As an administrative matter, however, such an approach would be massively complex and likely unworkable.

The approach incorporated in the Lieberman-Warner Bill to determine whether another nation has taken comparable action is to measure GHG emissions each year against a baseline. As an initial matter, nations like China or India could argue that emissions should be measured by geographical location of consumption, not production. China, after all, now produces half the world's cement and flat glass and a third of its steel; industry thus accounts for 71 percent of energy demand in China, as compared to 31 percent in Europe and 25 percent in the United States.[46] Leaving aside such normative questions, measuring each nation's emissions against a baseline also ignores that two nations may rationally achieve identical long-term reductions according to different annual emission patterns and from different sectors of the economy. The long-

43. *Canada-Automobiles*, paragraph 86.
44. Matsushita, Schoenbaum, and Mavroidis (2003, 147).
45. World Bank (2008, 24).
46. Rosen and Houser (2007).

term cumulative nature of climate change means that the marginal benefits of reducing GHG emissions vary little from year to year, while the costs might vary greatly. Thus, it may be economically efficient for a country to make fewer cuts in the short term and more in the long term, and annual measures of GHG emissions to determine comparable action would fail to allow for this temporal flexibility. Additionally, determining whether a nation had taken comparably effective measures by measuring GHG emissions reductions would fail to take into consideration the impact of land use changes and deforestation on climate change, which account for roughly one-fifth of GHG emissions.[47] It also ignores that some countries might increase (or decrease) emissions from a given baseline due to changes in population growth or GDP growth or other factors. For example, Russia and its former republics have lower emissions than they did in 1990 and thus can easily hit a target such as reducing emissions to or below 1990 levels.

Article XX Exceptions

On the basis of the WTO jurisprudence discussed above, there is reason to believe that a border adjustment that requires importers of carbon-intensive goods to purchase allowances at the U.S. market price for the carbon emitted in production might be found to violate the United States' most-favored-nation treatment obligations if applied only to countries that do not have comparably effective policies. As for the United States' national treatment obligations, a border adjustment that charged "like" products differently based on how much carbon was emitted in producing each product might be viewed as a prohibited PPM restriction. In that case, the border adjustment would then be permissible only if it satisfied one of the environmental exceptions in Article XX of the GATT and then, if it did, whether it was also consistent with the introductory paragraph ("chapeau") of Article XX.

The most relevant exceptions are found in Article XX(g) and XX(b). The exception in Article XX(g) applies to measures "relating to the conservation of exhaustible natural resources if such measures are made effective in conjunction with restrictions on domestic production or consumption." Article XX(b) provides an exception for measures "necessary to protect human, animal or plant life or health." Because the WTO has found that "relating to" is a lower standard to meet than "necessary to,"[48] this chapter focuses on the Article XX(g) exception.

---

47. According to the World Resources Institute (2008), in 2000, 23 percent of CO_2 emissions (18 percent of all GHG emissions) came from land use change and forestry.

48. See *US-Gasoline AB*, 14–19.

ARTICLE XX(G) EXCEPTION FOR "CONSERVATION OF EXHAUSTIBLE NAT-
URAL RESOURCES." There are three parts to the Article XX(g) analysis, two
of which should be satisfied without much difficulty. A low-carbon atmosphere,
necessary to avoid catastrophic climate change, should be viewed as an
"exhaustible natural resource"; although carbon only stays in the atmosphere
for around a hundred years, the WTO has previously found that clean air is a
resource capable of depletion even if it is renewable.[49] A border adjustment
would also be "made effective in conjunction with restrictions on domestic production or consumption" if it is part of an overall U.S. cap-and-trade bill.

A more difficult question is whether a border adjustment is "related to" the
goal of mitigating climate change. GATT panels have interpreted "relating to"
to mean "primarily aimed at" conservation.[50] In *US-Gasoline*, the Appellate
Body found the disputed measure satisfied XX(g) because it had a "substantial relationship" to the conservation of clean air.[51] In *US-Shrimp*, XX(g) was
satisfied because the import ban on shrimp harvested without devices to avoid
harming turtles while fishing demonstrated a "means and ends relationship"
that was "close and real" with the goal of protecting endangered turtles.[52] It is
less clear whether a border adjustment would satisfy the test of being primarily aimed at and substantially related to the goal of reducing GHG emissions
when estimates suggest the policy might do little to reduce leakage. Indeed, in
explaining why the disputed measures satisfied XX(g), the Appellate Body in
US-Gasoline noted that without baselines, the goal of reducing the level of air
pollution "would be substantially frustrated."[53] In *US-Shrimp*, the Appellate
Body accepted that turtle excluder devices "would be an effective tool for the
preservation of sea turtles."[54] It is harder to argue that the United States' goal
of mitigating climate change would be less "effective" or "substantially frustrated" without border adjustments on carbon-intensive imports from certain
countries.[55]

Conversely, the Appellate Body in those cases did not ask *how much* of an
impact the policy would have on protecting sea turtles, only whether it would
have that effect. Similarly, the fact that border adjustments for climate change
would have limited impact on total emissions should not necessarily count
against them. It is not relevant under XX(g) (as it is under XX(b)) that there

49. *US-Gasoline AB*, 14.
50. *US-Gasoline AB*, 18–19; GATT, *Panel Report on Canada: Measures Affecting Exports of
Unprocessed Herring and Salmon*, L/6268, BISD 35S/98, March 22, 1988, paragraphs 4.5–4.6.
51. *US-Gasoline AB*, 19.
52. *US-Shrimp AB*, 141.
53. *US-Gasoline AB*, 19.
54. *US-Shrimp AB*, 141.
55. *US-Shrimp AB*, 141.

may be more effective or fewer trade restrictive options available. WTO member governments retain "a large measure of autonomy to determine their own policies on the environment."[56] Moreover, the Appellate Body has previously explained that XX(g) must be "read . . . in light of contemporary concerns of the community of nations about the protection . . . of the environment,"[57] and few issues are of such universal concern at present as climate change.

ARTICLE XX CHAPEAU. Even if a border adjustment satisfies XX(g), it must also be justified under the "chapeau," or opening clause, to Article XX, designed to prevent provisions that are arbitrary, discriminatory, or protectionist. The chapeau requires that "measures are not applied in a manner which would constitute a means of arbitrary or unjustifiable discrimination between countries where the same conditions prevail, or a disguised restriction on international trade." Broadly speaking, the purpose of the chapeau is to prevent the "abuse of the exceptions" in Article XX,[58] and ensure that they are "exercised in good faith to protect interests considered legitimate under Article XX, not as a means to circumvent one Member's obligations towards other WTO Members."[59] The chapeau thus embodies the recognition by WTO members of the need to maintain a balance between the right to invoke an Article XX exception and the substantive rights under the GATT.[60] Several of the leading environmental cases under the WTO have turned on the standards set forth in the chapeau. In *US-Gasoline*, *US-Shrimp*, and most recently *Brazil-Tyres*, for example, the Appellate Body found the offending measure to be provisionally justified by one of the environmental paragraphs of Article XX—either XX(g) or XX(b)—but then found the measure violated the chapeau of Article XX.

The Appellate Body has explained that whether the application of a measure violates the chapeau "should focus on the cause or rationale given for the discrimination."[61] As a theoretical matter, a full border adjustment levied based on the carbon-intensity of an import has a plausible rationale consistent with XX(g), namely, to minimize carbon leakage that can undermine the effectiveness of a U.S. cap-and-trade system in lowering GHG emissions. Moreover, excluding nations with comparably effective climate policies from the scheme also has a defensible environmental rationale, which is that those nations are already taking measures to reduce carbon emissions and thus leakage to them should be of minimal concern.

56. *US-Gasoline AB*, 30.
57. *US-Shrimp AB*, 129.
58. *US-Gasoline AB*, 22.
59. WTO, *Appellate Body Report on Brazil: Measures Affecting Imports of Retreaded Tyres*, WT/DS332/AB/R, December 3, 2007 [hereafter *Brazil-Tyres AB*], paragraph 215.
60. *US-Shrimp AB*, 156.
61. *Brazil-Tyres AB*, 246.

At the same time, the chapeau addresses the "detailed operating provisions" of the measure at issue and how it is "actually applied."[62] When consideration is given to the administrative complexities and likely implementation, there are at least five possible reasons why a border adjustment, depending on how it is designed, may violate the chapeau.

First, as discussed above, a border adjustment on carbon-intensive manufactured goods from countries that have not taken comparably effective action to address climate change, as commonly proposed today, would do little to reduce overall leakage and have little environmental benefit. For those carbon-intensive sectors that face competition from Annex II countries, however, such border adjustments may help mitigate the adverse effects of a carbon price.[63] Yet there is no exception in Article XX for preserving the health of U.S. firms, only the environment, and thus the ostensible purpose for the offending measure must be a "rational connection" to an objective of one of the Article XX paragraphs.[64] One leading scholar explains it this way: "It is one thing for the United States to demand that the shrimp it imports be caught in a turtle-safe way so as to safeguard turtles. Yet it is an entirely different matter to seek to 'level the playing field' by insisting that foreign producers use the same production practices as U.S. shrimpers so as to offset any regulatory cost differences between domestic and foreign producers. This latter motivation should not be shielded by GATT Article XX."[65] A WTO panel, balancing the rights of a member to invoke an Article XX derogation and the rights of other members,[66] may thus find border adjustments to be a form of stealth protectionism given the larger impact on protecting certain U.S. firms than on reducing overall GHG emissions. Indeed, the Appellate Body recently confirmed that it is acceptable to take into account as a relevant factor the *effects* of discrimination, not just the cause or rationale of the discrimination.[67]

Second, the measure likely will have to permit importers to demonstrate how much carbon they emitted individually and pay for allowances on that basis. The current provision in the Lieberman-Warner Bill, for example, does not do that. Rather, it prescribes a particular formula used to determine the allowance requirement for a category of covered goods in a covered foreign country.[68]

62. *US-Shrimp AB*, 160.
63. See *US-Shrimp AB*, notes 4–18. As Charnovitz (2002, 148) has noted in the context of climate change policies, "while one can easily see a competitiveness rationale to use a border tax adjustment, it is difficult to visualize a valid environmental reason under GATT Article XX in support of a border adjustment."
64. *Brazil-Tyres AB*, 227.
65. Charnovitz (2002, 106).
66. See *US-Shrimp AB*, 156.
67. *Brazil-Tyres AB*, 230.
68. ACSA, Title VI, Sec 6001(d).

Manufacturers in that category would thus have the same allowance requirement regardless how much carbon each actually emitted in production. Such a provision may be ruled arbitrary and unjustifiable discrimination, just as the baseline establishment method in *US-Gasoline* was found to be because domestic refiners were permitted to establish an individual baseline while foreign refiners had to accept the Environmental Protection Agency's statutory baseline. Moreover, use of a nationwide calculus for carbon content actually weakens the ostensible climate benefits of the border adjustment. Foreign firms will have little incentive to reduce the carbon footprint of their products if doing so would not change the way their products are treated at the border (though the incentive for government action may still exist).

Third, the United States cannot require an exporting country to implement a market mechanism, but it must allow flexibility for nations to pursue other approaches "comparable in effectiveness."[69] The Appellate Body has interpreted "arbitrary or unjustifiable discrimination" to preclude requiring essentially the same program as the United States puts in place to address climate change.[70] As a practical matter, this may significantly mute the impact of border adjustments. As discussed above, it is difficult to judge the efficacy of climate change policies in the short term, and thus other nations might argue that a variety of policies—from command-and-control regulations to the sort of voluntary targets adopted in Japan—should be viewed as comparably effective.

Fourth, the U.S. program must take "into consideration different conditions which may occur" in different countries.[71] Failure to do so may constitute "arbitrary discrimination," according to the Appellate Body.[72] In that regard, the WTO might consider the relevance of developed countries' greater historical responsibility for cumulative carbon emissions and higher current emissions per capita. In that case, there is a possibility the WTO would find that even a border adjustment applied equally to domestic and imported goods is noncompliant.

Fifth and finally, the Appellate Body's interpretation of the chapeau requires that before imposing a border adjustment, the United States must engage in "serious, across-the-board negotiations" with other nations that might be subject to the border adjustments.[73] Although the United States must make

---

69. *US-Shrimp*, Article 21.5 AB, 137–44. It is also worth noting that a trade mechanism that requires comparability in form as well as burden runs counter to the Framework Convention on Climate Change and the Kyoto Protocol, which provides discretion to countries on how they implement their climate change policy goals.

70. *US-Shrimp*, Article 21.5 AB, 144.

71. *US-Shrimp*, Article 21.5 AB, 164.

72. *US-Shrimp*, Article 21.5 AB, 164, 165, 177.

73. *US-Shrimp*, Article 21.5 AB, 166-71.

good-faith efforts to reach agreement, failure to do so will not constitute discrimination.[74] If the United States were not to undertake such negotiations, or the WTO were to view the efforts as insufficiently "serious," that might lead to a finding that the border adjustment is arbitrary and discriminatory.[75]

The Alternative of Free Allocation

Questions about the WTO legality of border adjustments, combined with concerns about tit-for-tat trade retaliation and their small impact on reducing leakage, may caution against their use. Yet, as a practical matter, some measures may be necessary in a U.S. cap-and-trade law to address potential adverse effects on domestic manufacturers. In response to this political imperative, some have proposed compensating adversely affected sectors in the United States through the free allocation of emission allowances.[76] Though the topic of free versus auctioned allocation has attracted considerable attention as a policy matter, to date little attention has been given to the question whether free allocation, too, might be noncompliant with WTO law as an illegal subsidy. As discussed below, to the extent it *is* compliant, that may only be precisely for the same reasons it is *not* likely to be good policy.

Under the WTO Agreement on Subsidies and Countervailing Measures, free allocation would be a subsidy subject to remedies under the agreement if it (1) were a "financial contribution" by the government, (2) conferred a "benefit," and (3) were "specific" to certain industries or sectors.[77] If these elements are satisfied, the subsidy may be inconsistent with WTO law if it also causes adverse effects to other WTO members.[78]

74. *US-Shrimp*, Article 21.5 AB, 115–24.

75. Efforts to undertake "serious, across-the-board negotiations" before the imposition of a border adjustment may be complicated by the most recent Senate climate change bill's reduction from eight years to two years the delay between the start of the domestic cap-and-trade program and the start of the international allowance program. Sec. 1315.

76. See, for example, Pew Center on Global Climate Change (2008).

77. WTO, "Agreement on Subsidies and Countervailing Measures," Articles 1.1, 1.2.

78. WTO, "Agreement on Subsidies and Countervailing Measures," Article 5. In addition to being an "actionable" subsidy if it causes adverse effects, a claim might also be made that it constitutes a "prohibited" export-contingent subsidy, which is forbidden per se. World Trade Organization, "Agreement on Subsidies and Countervailing Measures," Article 3. Though a subsidy may be prohibited if it is contingent de facto or de jure on export (*Appellate Body Report on Canada: Measures Affecting the Export of Civilian Aircraft*, WT/DS70/AB/R, August 2, 1999, paragraph 167), export orientation alone is not enough; the subsidy must be "in fact tied to actual or anticipated exportation or export earnings." World Trade Organization, "Agreement on Subsidies and Countervailing Measures," n. 4. Free allocation to carbon-intensive industries is unlikely to meet that test.

First, free allocation of allowances should be considered a financial contribution, which is defined, among other ways, as the "direct transfer of funds, such as grants, loans and equity infusions."[79] Though it might be argued that free allowances do not constitute the "direct transfer of funds," they are "functionally equivalent to distributing cash," according to the Congressional Budget Office (CBO), because allowances can be sold for monetary value in a liquid secondary market, created and enforced by the government.[80] Indeed, that is why the recipients of allowances need not be the same entities regulated by the cap. In recognition of the financial value of free allowances, the CBO recently scored free allocation as both a revenue increase and outlay increase. As the then–CBO director, Peter Orszag, explained, "Distributing allowances at no charge to specific firms or individuals is, in effect, equivalent to collecting revenue from an auction of the allowances and then distributing the auction proceeds to those firms or individuals."[81]

Second, if free allowances are a "direct transfer of funds" qualifying as a financial contribution by the government, it should be readily agreed that they also confer a benefit. "If a government gives a sum of money to a company, it seems clear that this financial contribution would generally confer a benefit."[82]

Third, a subsidy is "specific" when it is provided to a specific industry or enterprise but not when it is widely available within an economy.[83] To the extent free allowances are targeted at specifically defined sectors adversely affected by a carbon price, they would likely be considered specific. Conversely, if all allowances were distributed based on objective criteria, like historical emissions, it would be harder to prove that the subsidies were specific.[84] Even if a subsidy is de jure nonspecific, however, it can still be de facto specific if, for example, certain enterprises benefit disproportionately.[85]

Finally, for the program to be an "actionable" subsidy, it must cause "adverse effects" to the interests of another WTO member.[86] The most likely way in which free allocation may be found to do so would be if it caused "serious prejudice," notably because the "subsidy displaces or impedes imports of a like product of another Member in the market of the subsidizing Member."[87]

79. WTO, "Agreement on Subsidies and Countervailing Measures," Article 1.1(a)(1)(i).
80. Congressional Budget Office (2007a).
81. Orszag (2008).
82. Van den Bossche (2005, 557).
83. WTO, "Agreement on Subsidies and Countervailing Measures," Article 2.
84. WTO, "Agreement on Subsidies and Countervailing Measures," n. 2.
85. WTO, "Agreement on Subsidies and Countervailing Measures," Article 2.1(c).
86. WTO, "Agreement on Subsidies and Countervailing Measures," Article 5.
87. WTO, "Agreement on Subsidies and Countervailing Measures," Article 6.3(a).

Although free allocation may appear, at first glance, to harm other WTO members by reducing costs for domestic producers, in fact free allocation should not change a firm's pricing and output decisions, and thus foreign firms should not see their sales reduced by artificially suppressed prices for U.S. goods. As discussed above, free allocation of allowances does *not* exempt firms from the carbon price signal created by a cap-and-trade system. Rather, it is a transfer of resources from the government to the recipients. Even if firms receive allowances for free, they will still pass along the opportunity cost of using those allowances to their customers in the form of higher prices.[88] Indeed, in Europe, which gave allowances away for free, consumers still saw electricity prices rise and fall with the market value of allowances, while firms reaped windfall profits. As prices increase, demand falls, and the firm's output is reduced accordingly.[89] How much output is reduced should not differ depending on whether allowances are auctioned or freely distributed. Firms set prices based on market forces, such as marginal costs and demand, that do not change even if firms receive a cash transfer from the government. This economic effect of free allocation is particularly important to recognize in considering how to protect U.S. industry, because free allocation will increase firms' profits, which ultimately accrue to shareholders, but will *not* prevent production declines and concomitant job losses in affected sectors.[90] Free allocation thus may not adversely affect other WTO members or be illegal under WTO law—though precisely because it would be *ineffective* in protecting U.S. industries or workers (though it would compensate shareholders). Even if output and pricing decisions are unchanged, however, WTO members may claim they suffered "serious prejudice" if free allocation provides firms with resources to invest in research and development (R&D) or new products, prevents exit from the market, or has other indirect benefits.

It is worth noting that whether free allocation has adverse effects on importers may depend on the formula used for the allocation and updating of allowances. For example, "output-based allocation," which some analysts have proposed,[91] *would* affect the pricing and output decisions of firms.[92] This approach would

---

88. See Congressional Budget Office (2007b, 1; 2003).

89. For estimates of output reductions from a carbon charge, see Morgenstern and others (2007).

90. See, for example, Orszag (2008): "Because the additional profits from the allowances' value would not depend on how much a company produced, such profits would be unlikely to prevent the declines in production and resulting job losses that the price increases (and resulting drop in demand) would engender."

91. Fischer and Fox (2004); Fischer, Hoffmann, and Yoshino (2002). Under output-based allocation, a certain number of allowances are allocated to certain sectors, and within each sector each firm would receive a number of permits proportional to its share of the sector's output.

92. In addition to the legal issues, by not allowing the carbon price signal to be passed through

be the functional equivalent of auctioning off allowances and then using the revenue to subsidize production. Indeed, some proposals on Capitol Hill make this explicit, rebating firms in cash for the average carbon costs (from the purchase of allowances and higher electricity prices) associated with output in their sector—thus preserving some incentive to reduce energy intensity relative to the sector average. With output-based allocation or rebating, output would thus be reduced less than would be the case absent the production subsidy—thereby giving rise to claims of "serious prejudice" by other WTO members.[93] In short, the more effective free allocation is in protecting employment and output in adversely affected sectors, the more likely it may be to violate WTO law. If free allocation were found to be a subsidy inconsistent with WTO law, the adversely affected WTO member might seek the right to retaliate against U.S. products imported to that country if the free allocation were not removed.[94]

In response to a claim by a WTO member that free allocation constituted an illegal subsidy, the United States might respond in its defense that the WTO Agreement on Subsidies and Countervailing Measures permits certain environmental adaptation subsidies, and that free allocation of permits falls within that provision as a subsidy to help manufacturers adjust to the effects of climate policy.[95] Although this provision expired in 2000, the United States might

in the form of higher prices, output-based allocation is a less efficient way to reduce emissions because it reduces conservation incentives and instead increases reliance on lowering energy intensity. Conversely, its proponents argue that output-based allocation would reduce leakage, mitigate adverse effects on employment in the affected sector, and reduce negative effects on the real wage by mitigating price increases of energy-intensive goods.

93. In its defense, the United States might argue that output-based allocation is not a *net* subsidy once the impact of the cap-and-trade system is considered, so importers are not adversely affected relative to the business-as-usual scenario of no U.S. climate change regulation. In response, the complaining WTO member might make two points. First, it is theoretically possible for firms to receive a net subsidy under output-based allocation if the sector-level allocations are based on historical emissions shares but a given sector or subsector does relatively more than others to reduce emissions. Second, the member might argue that the measure being disputed as an actionable subsidy is not the U.S. climate regulation (including the allocation method), but rather output-based allocation itself. Under a cap-and-trade system, the economy as a whole would face higher costs from the domestic climate regulation, but output-based allocation would be a way to offset part of those costs for a certain subset of industries—in a way that induces higher output and thus might harm importers. In that way, exempting certain sectors from regulatory costs (and concomitant GHG reductions) that would otherwise exist may be viewed by the WTO as similar to "forgoing" revenue "otherwise due," which is a financial contribution under Article I of the "Agreement on Subsidies and Countervailing Measures." When a regulation applies to the entire economy, but only certain sectors receive a subsidy that mutes the regulation's impact, the WTO may find that the benchmark against which to evaluate whether a member has been adversely affected is the absence of the disputed method of allocation, not the absence of the entire regulatory program.

94. WTO, "Agreement on Subsidies and Countervailing Measures," Articles 10–11. In addition, the affected foreign industry might consider bringing a countervailing duty case under its domestic trade laws.

95. WTO, "Agreement on Subsidies and Countervailing Measures," Article 8.2(c).

argue that the environmental exceptions to which negotiators agreed in 1995 are still justified today. Yet even if the WTO were to agree that the environmental exceptions existed notwithstanding their expiration, free allocation of permits is unlikely to fall within the narrow scope of the exceptions. The subsidy must provide "assistance to promote adaptation of existing facilities to new environmental requirements imposed by law and/or regulations which result in greater constraints and financial burden on firms,"[96] but free allocation is a cash transfer that firms may use to "green" their facilities or may invest in other more profitable ventures or may distribute as profits to shareholders. Further, environmental adaptation subsidies must be a one-time subsidy,[97] whereas free allocation is likely to take place annually. The subsidies must be limited to 20 percent of the adaptation costs, yet most proposals to compensate carbon-intensive sectors propose much larger subsidies.[98] The subsidies also must be "directly linked to and proportionate to a firm's planned reduction of nuisances and pollution," but, as noted, free allocation is effectively a cash transfer that creates no obligation to invest in cleaner production. Fundamentally, the problem is that the provision for environmental adaptation subsidies was written to address the costs of complying with command-and-control regulations, like improved pollution and efficiency standards, but is inapposite to a market mechanism like a carbon price created by a cap-and-trade regime.

Conclusion

As explained in this chapter, the consistency of border adjustments with WTO law is in doubt and may come down to whether the WTO panel finds the measure to be a genuine effort to protect the environment or a form of stealth protectionism. The alternative of free allocation may be more WTO compliant, though only because it should not boost output from affected industries. Free allocation of allowances, however, does represent a large cash transfer to domestic firms, while doing little to reduce job loss in affected sectors. To some

96. WTO, "Agreement on Subsidies and Countervailing Measures," Article 8.2(c).
97. WTO, "Agreement on Subsidies and Countervailing Measures," Article 8.2(c)(i).
98. Several studies, for example, found that free distribution of around 20 percent of allowances would be sufficient to compensate primary energy producers and electric power generators. See Smith, Ross, and Montgomery (2002); Burtraw and Palmer (2006); and Burtraw and others (2002). A subsidy of 20 percent of total costs would thus require free allocation of only 4 percent of allowances, whereas Lieberman-Warner proposed freely allocating about 30 percent of all allowances each year in its first ten years to owners of manufacturing, fossil fuel–fired electricity generators, petroleum refiners, and natural gas processors.

extent, such job loss is an inevitable consequence of reduced demand for carbon-intensive goods, which is a key purpose of a carbon price signal. Rather than benefit shareholders with free allocation, a better approach would be to auction allowances and use some of the revenue to assist workers transition—perhaps to "greener" jobs that new investment incentives will create, through a sort of "carbon price displacement assistance" program. In addition, auction revenue can offset the distributional impact of a carbon price through progressive tax policy, reduce other distortionary taxes, and permit greater investment in environmental R&D. In short, the expected environmental benefit of border adjustments for carbon-intensive manufactured goods is likely to be quite small compared with the trade and WTO risks they pose, and any adverse effects on employment in affected carbon-intensive industries due to the lack of a border adjustment can be mitigated through the well-targeted use of allowance auction revenue. It is also important to point out that the costs to industry of complying with climate policy in the first place can be minimized by using a cost-efficient market mechanism, such as a cap-and-trade system or carbon tax.

Ultimately, all the problems and challenges associated with measures to address competitiveness and leakage reinforce the truly global nature of climate change and the limited ability of any one country to address it unilaterally. International engagement is thus critical to mitigate climate change, and a new post–Kyoto Protocol international architecture will be needed in this regard.[99] Achieving this goal is complicated, however, by considerations of economic efficiency, which requires low-cost abatement in the developing world, and distributional equity, which demands action from rich nations historically responsible for emitting GHGs. Until a truly international approach is adopted, it is critical that the United States, at long last, show real leadership and adopt serious unilateral measures to curb GHG emissions, while high-income countries take collective steps to assist the rest of the world in reducing its emissions.

99. See, generally, Aldy and Stavins (2007); and Stern and William Antholis (2007).

Comment

Comment by Andrew W. Shoyer

In chapter 2, Jason Bordoff discusses the risk that any "border adjustment" included as a component of U.S. climate change legislation would be ruled noncompliant with World Trade Organization (WTO) requirements. As a threshold matter, the use of the term "border adjustment" is unfortunate and somewhat misleading in this context, because it has no precise meaning in international trade law, in contrast to the term "border tax adjustment."[1]

However, at least one proposal that Bordoff would consider a border adjustment—the international reserve allowance program proposed initially by the chief executive officers of the International Brotherhood of Electrical Workers and American Electric Power,[2] and reflected most recently in Title XIII, Subtitle A, of Senator Barbara Boxer's "Manager's Amendment" to S 3036[3]—was designed to be WTO compliant. It would apply to imports, such as carbon-intensive products, the same type of measures that would apply under a cap-and-trade program to U.S. producers. And it would apply these measures to imports from countries that are major emitters of greenhouse gases (GHGs) and that had failed to take action comparable in impact to that of the United States to reduce these emissions. But the ultimate goal of the program is that the import measures never take effect—that the leverage offered to U.S. negotiators equipped with the credible threat of WTO-compliant measures will induce large emitters to take effective action promptly on their own and through international negotiations to limit emissions.

The underlying assumption of the international reserve allowance program is that the United States would enact mandatory GHG emissions reductions—

1. See, for example, "Report of the Working Party on Border Tax Adjustments, Adopted by the GATT 1947 Contracting Parties on December 2, 1970," L/3436, BISD 18S/97.
2. Morris and Hill (2007).
3. Senate Amendment 4825, *Cong. Rec.* S5049, S5091–95, June 4, 2008 [hereafter Manager's Amendment].

in the form of a cap-and-trade system—before it and other countries concluded and put in place a post–Kyoto Protocol agreement. As described in the Manager's Amendment to S 3036, the U.S. government would, upon entry into force of cap-and-trade legislation, notify all other countries of the U.S. annual GHG emissions caps.[4] The U.S. government would then negotiate with GHG-emitting countries to secure internationally agreed-on disciplines on emissions.[5] The international reserve allowance program is targeted at exports from those countries that do not take "comparable action" to that taken by the United States.[6] The U.S. government would begin to measure on an annual basis the reduction of emissions under the U.S. cap and use those data to determine whether and to what extent other countries had taken "comparable action." The determination of whether a country had taken such action would be based, therefore, on the impact of a country's regulation on emissions rather than the precise form of the regulatory program used to achieve those effects.[7] The U.S. government would focus its determination on those countries that contribute most to global emissions—least developed countries and countries with less than a de minimis volume of emissions would be excluded from the international reserve allowance requirements.[8]

If the U.S. government determined that a country did not take comparable action, then an importer of certain goods (for example, carbon-intensive goods, such as steel and aluminum) from that country would be required to provide allowances to the U.S. government corresponding to the GHGs emitted when the imported goods were produced in the country of origin.[9] The U.S. government would use an adjustment factor in setting the number of allowances required for imported goods.[10] This adjustment factor would reflect the portion of allowances that domestic producers receive at no cost in relation to the allowances that domestic producers procure by auction.[11] The adjustment factor would also reflect the conditions prevailing in different countries. The U.S. government would permit importers to satisfy their obligations using allowances (and credits) generated under the cap-and-trade systems of other countries, or using "international reserve" allowances sold by the U.S. government at the

---

4. Manager's Amendment, §1303(c).

5. Manager's Amendment, §1303.

6. Senate Amendment 4825, sections 1301(4), 1305, *Cong. Rec.* S5049, S5091–92, June 4, 2008.

7. Senate Amendment 4825, sections 1301(4), 1305, *Cong. Rec.* S5049, S5091–92, June 4, 2008.

8. Manager's Amendment, §1306(b)(2).

9. Manager's Amendment, §1306(c).

10. Manager's Amendment, §1306(d).

11. Manager's Amendment, §1306(d)(4).

prevailing price of U.S. cap-and-trade allowances.[12] There would be no limit to the number of international reserve allowances made available for purchase by importers, and these international reserve allowances could not be used by U.S.-regulated entities to satisfy their obligations under the U.S. cap-and-trade system.[13]

In chapter 2, Bordoff is careful not to draw conclusions about the legality under the WTO of any particular proposal. It is noteworthy, however, as described below, that the international reserve allowance program is designed to fall within the exception of the WTO General Agreement on Tariffs and Trade (GATT) for measures related to the conservation of exhaustible natural resources, by

—securing a close "ends-means" relationship with the overall environmental objectives of the cap-and-trade program;

—implementing measures in conjunction with limitations on U.S. production, in an "evenhanded" fashion, so that foreign goods are not treated worse than domestic goods;

—adjusting import requirements to take into account different conditions among countries;

—allowing time for good-faith negotiating efforts with all affected countries; and

—allowing time to measure U.S. emissions reductions before requiring importers to secure allowances.

Each of these elements is discussed in turn.

The program provides a real solution to the conservation objective of reducing GHG emissions. GATT Article XX(g) provides a general exception to the GATT's substantive obligations only for those government measures that are "primarily aimed at" the conservation of exhaustible natural resources. In its *US-Shrimp* decision,[14] the WTO Appellate Body recognized that a government measure was primarily aimed at the conservation of an exhaustible natural resource if "a close and genuine relationship of ends and means" exists between the measure and the conservation objective.[15] Under the international reserve allowance program, importers could meet the requirements by providing allowances from recognized cap-and-trade programs outside the United States,

12. Manager's Amendment, §1306(e).
13. Manager's Amendment, §1306(a)(7).
14. See WTO, *Appellate Body Report on United States: Import Prohibition of Certain Shrimp and Shrimp Products*, WT/DS58/AB/R, November 6, 1998, DSR 1998:VII, 2755 [hereafter *US-Shrimp AB*]; WTO, *Appellate Body Report on United States: Import Prohibition of Certain Shrimp and Shrimp Products—Recourse to Article 21.5 of the DSU by Malaysia*, WT/DS58/AB/RW, November 21, 2001, DSR 2001:XIII, 6481 [hereafter *US-Shrimp 21.5 Proceedings AB*].
15. *US-Shrimp AB*, paragraph 136.

or by securing international reserve allowances from the U.S. government, both of which are designed to reflect actual GHG emissions abroad. In contrast, a carbon tax on imports would have no direct relationship to the reduction of emissions abroad.

The program, which would place restrictions on the importation of certain foreign products, is implemented in parallel with restrictions on domestic production. GATT Article XX(g) applies "if such measures are made effective in conjunction with restrictions on domestic production or consumption"— language that the WTO Appellate Body has interpreted as requiring "even-handedness."[16] In other words, as explained by the Appellate Body in *US-Gasoline*,[17] restrictions on imported products must be "promulgated or brought into effect together with restrictions on domestic production or consumption of natural resources."[18] However, the Appellate Body also made clear in its *US-Gasoline* decision that GATT Article XX(g) does not require "identical treatment of domestic and imported products."[19]

The program is structured to take into consideration the different conditions that may exist in affected exporting countries. According to the Appellate Body in *US-Shrimp*, the chapeau of GATT Article XX requires that a government measure "be designed in such a manner that there is sufficient flexibility to take into account the specific conditions prevailing in any exporting Member."[20] In contrast, a single carbon-intensity standard for all products in a particular sector could not meet this requirement. In its *US-Shrimp* decision, the Appellate Body found unacceptable government measures that "require other [WTO] Members to adopt essentially the same comprehensive regulatory program, to achieve a certain policy goal, as that in force within that Member's territory, without taking into consideration different conditions which may occur in the territories of those other Members."[21] Moreover, the Appellate Body has found a government measure that "condition[s] market access on the adoption of a programme comparable in effectiveness" (versus the same program) satisfies the chapeau's requirements because the measure permits sufficient flexibility in its application.[22]

16. *US-Gasoline AB*, 20–21; *US-Shrimp AB*, paragraphs 144–45.

17. WTO, *Appellate Body Report on United States: Standards for Reformulated and Conventional Gasoline*, WT/DS2/AB/R, May 20, 1996, DSR 1996:I, 3 [hereafter *US-Gasoline AB*]; WTO, *Panel Report on United States: Standards for Reformulated and Conventional Gasoline*, WT/DS2/R, May 20, 1996, as modified by WTO, *Appellate Body Report*, WT/DS2/AB/R, DSR 1996:I, 29 [hereafter *US-Gasoline Panel*].

18. *US-Gasoline AB*, 18.

19. *US-Gasoline AB*, 21.

20. *US-Shrimp 21.5 Proceedings AB*, paragraph 149.

21. *US-Shrimp AB*, paragraph 164.

22. *US-Shrimp 21.5 Proceedings AB*, paragraph 144.

The program provides sufficient time for the U.S. government to engage in serious negotiations with all affected countries to curb GHG emissions before the international allowance requirement would enter into effect. The Appellate Body rejected the government measure at issue in *US-Shrimp* in part because of "the failure of the United States to engage the appellees, as well as other Members exporting shrimp to the United States, in serious, across-the-board negotiations with the objective of concluding bilateral or multilateral agreements for the protection and conservation of sea turtles, before enforcing the import prohibition against the shrimp exports of those other Members."[23] Moreover, in *US-Shrimp*, the Appellate Body found a violation of the anti-abuse provisions in the chapeau because "the United States negotiated seriously with some, but not with other Members" that were similarly situated.[24]

To be clear, the Appellate Body has not interpreted GATT Article XX as requiring that a WTO member government negotiate with other governments before it imposes an environmental measure. Rather, the chapeau of Article XX requires nondiscrimination, so that if a WTO member government chooses to negotiate with some countries that would be affected by a measure, it must negotiate with all such countries.

The United States is already negotiating climate issues with other nations, and it will discuss the application of the international allowance provision with some of the other nations that are affected by it. To meet the GATT Article XX criterion, therefore, the United States will be obligated to negotiate with all the countries to which the provision will be applied (but not those exempted from the measure), because it will be negotiating with some of them.

The United States is not required to conclude negotiations—only to make serious, good-faith efforts with all (approximately thirty) affected countries. The negotiations could commence immediately upon passage of the legislation and its enactment into law. Thus, the requirement to negotiate does not affect the date on which the allowance requirement would be imposed on imports from affected countries.

The program imposes the international allowance requirement on imports after the cap-and-trade requirements would apply to domestic production, so that importers are provided in advance with the standard of comparability of action. In *US-Tuna I*,[25] the GATT 1947 Panel noted (in an unadopted report) that because the United States had "linked the maximum incidental dolphin-taking rate which Mexico had to meet during a particular period in order to be

23. *US-Shrimp AB*, paragraph 166.
24. *US-Shrimp AB*, paragraph 172.
25. GATT, *Panel Report on United States: Restrictions on Imports of Tuna*, DS21/R, GATT BISD 39S/155 [hereafter *Tuna I*], circulated September 3, 1991; not adopted.

able to export tuna to the United States to the taking rate actually recorded for United States fisherman during the same period," the "Mexican authorities could not know whether, at a given point of time, their conservation policies conformed to the United States conservation standards."[26] The panel concluded that "a limitation on trade based on such unpredictable conditions could not be regarded as being primarily aimed at the conservation of dolphins."[27]

Under the international reserve allowance program, the U.S. president would be required immediately upon enactment to notify other countries of annual U.S. emission caps specified in the legislation, providing a predictable standard several years in advance with respect to which foreign countries would adjust their GHG emissions regulations. But the allowance requirement would be applied on imports only after the U.S. government measured emissions reduction in the United States and provided that standard of "comparability" to producers in and importers from affected countries. Under WTO jurisprudence, the United States must apply the measure to affected countries in an "evenhanded" manner as compared with the manner in which it is applied to U.S. production or consumption.[28] If the United States requires concrete verification and measurable results in exporting countries, it will be difficult for the United States to justify not doing so with respect to the results achieved domestically under the cap. Conversely, if the United States were to apply the allowance requirement on imports immediately upon the application of the cap inside the United States, without any measurement or verified results of GHG emissions reductions, then "evenhandedness" would appear to require the United States to treat affected foreign countries in a similar fashion—without any measurement or verification of GHG emissions abroad.

26. *Tuna I*, paragraph 5.28.
27. *Tuna I*, paragraph 5.28.
28. *US-Gasoline AB*, 20–21; *US-Shrimp AB*, paragraphs 144–45.

References

Aldy, Joseph E., and William A. Pizer. 2008. *Competitiveness Impacts of Climate Change Mitigation Policies.* RFF Discussion Paper 08-21. Washington: Resources for the Future.

Aldy, Joseph E., and Robert N. Stavins. 2007. *Architectures for Agreement: Addressing Global Climate Change in the Post-Kyoto World.* Washington: Resources for the Future.

Bhagwati, Jagdish, and Petros C. Mavroidis. 2007. "Is Action against US Exports for Failure to Sign Kyoto Protocol WTO-Legal?" *World Trade Review, 6:* 299–310.

Bradford, Scott C., Paul L. E. Grieco, and Gary Clyde Hufbauer. 2005. "The Payoff to America from Global Integration." In *The United States and the World Economy: Foreign Economic Policy for the Next Decade*, ed. C. Fred Bergsten. Washington: Peterson Institute for International Economics.

Burtraw, Dallas, and Karen Palmer. 2006. "Compensation Rules for Climate Policy in the Electricity Sector." Paper presented at National Bureau of Economic Research Summer Institute, Workshop on Public Policy and the Environment, Cambridge, Mass.

Burtraw, Dallas, Karen Palmer, Ranjit Bharvirkar, and Anthony Paul. 2002. "The Effect on Asset Values of the Allocation of Carbon Dioxide Emission Allowances." *Electricity Journal* 15, no. 5: 51–62.

Charnovitz, Steven. 2002. "The Law of Environmental 'PPMs' in the WTO: Debunking the Myth of Illegality." *Yale Journal of International Law* 27, no. 59: 97.

———. 2003. *Trade and Climate: Potential Conflict and Synergies in Beyond Kyoto: Advancing the International Effort against Climate Change.* Washington: Pew Center on Global Climate Change.

Congressional Budget Office. 2003. *Shifting the Burden of a Cap-and-Trade Program.* Washington.

———. 2007a. "Cost Estimate, S. 2191: America's Climate Security Act of 2007" (www.cbo.gov/ftpdocs/91xx/doc9121/s2191_EPW_Amendment.pdf [March 2009]).

———. 2007b. *Trade-Offs in Allocating Allowances for CO_2 Emissions.* Washington.

DOE (U.S. Department of Energy). 2008. *International Energy Outlook 2008.* Report DOE/EIA-0484(2008).

Fischer, Carolyn, and Alan Fox. 2004. *Output-Based Allocations of Emissions Permits: Efficiency and Distributional Effects in a General Equilibrium Setting with Taxes and Trade.* RFF Discussion Paper 04-37. Washington: Resources for the Future.

Fischer, Carolyn, Sandra Hoffmann, and Yutaka Yoshino. 2002. *Multilateral Trade Agreements and Market-Based Environmental Policies.* RFF Discussion Paper 02-28. Washington: Resources for the Future.

Furman, Jason, Jason Bordoff, Manasi Deshpande, and Pascal Noel. 2007. *An Economic Strategy to Address Climate Change and Promote Energy Security.* Hamilton Project Strategy Paper. Brookings.

Houser, Trevor, Rob Bradley, Britt Childs, Jacob Werksman, and Robert Heilmayr. 2008. *Leveling the Carbon Playing Field: International Competition and US Climate Policy Design.* Washington: Peterson Institute for International Economics.

Hudec, Robert E. 2000. "The Product-Process Doctrine in GATT/WTO Jurisprudence." *In New Directions in International Economic Law: Essays in Honour of John H. Jackson*. ed., Marco Bronckers and Reinhard Quick. Boston: Kluwer.

Intergovernmental Panel on Climate Change. 2001. *Climate Change 2001: Mitigation: Summary for Policymakers*. Geneva.

Ismer, R., and K. Neuhoff. 2004. *Border Tax Adjustments: A Feasible Way to Address Nonparticipation in Emission Trading*. Cambridge Working Paper in Economics 0409, Cambridge University (http://ideas.repec.org/p/cam/camdae/0409.html [February 2009]).

Kopp, Raymond J. 2007. *Allowance Allocation*. Washington: Resources for the Future.

Krugman, Paul. 2008. "Trade and Wages Reconsidered." *Brookings Papers on Economic Activity*, ed. Douglas Elmendorf, N. Gregory Mankiw, and Lawrence Summers. Spring, 103–54.

Matsushita, Mitsuo, Thomas J. Schoenbaum, and Petros C. Mavroidis. 2003. *The World Trade Organization: Law, Practice, and Policy*. Oxford University Press.

McKibbin, Warwick J., M. Ross, R. Shackleton, and P. Wilcoxen. 1999. *Emissions Trading, Capital Flows and the Kyoto Protocol*. Discussion Paper in International Economics 144. Brookings.

Morgenstern, Richard D., Joseph E. Aldy, Evan M. Herrnstadt, Mun Ho, and William A. Pizer. 2007. *Competitiveness Impacts of Carbon Dioxide Pricing Policies on Manufacturing*. RFF Issue Brief 7. Washington: Resources for the Future.

Morris, Michael G., and Edwin D. Hill. 2007. "Commentary: Trade in the Key to Climate Change." *Energy Daily*, February 20.

Orszag, Peter R. 2008. "Implications of a Cap-and-Trade Program for Carbon Dioxide Emissions, Testimony before the Senate Finance Committee," April 24 (www.cbo.gov/ftpdocs/91xx/doc9134/04-24-Cap_Trade_Testimony.1.1.shtml [May 2008]).

Paltsev, Sergey V. 2001. "The Kyoto Protocol: Regional and Sectoral Contributions to the Carbon Leakage." *Energy Journal* 22, no. 4: 53–79.

Pauwelyn, Joost. 2007. *U.S. Federal Climate Policy and Competitiveness Concerns: The Limits and Options of International Trade Law*. Nicholas Institute for Environmental Policy Solutions Working Paper 0702. Durham, N.C.: Duke University.

Pew Center on Global Climate Change. 2008. "Response of the Pew Center on Global Climate Change to Climate Change Legislation Design White Paper: Competitiveness Concerns / Engaging Developing Countries" (www.pewclimate.org/docUploads/Pew%20Center%20on%t20Competitiveness-Developing%20Countries-FINAL.pdf [May 2008]).

Pew Research Center. 2008. "Obama's Image Slips, His Lead Over Clinton Disappears: Public Support for Free Trade Declines" (http://peoplepress.org/reports/ display.php3?PageID=1295 [May 2008]).

Rosen, Daniel H., and Trevor Houser. 2007. *China Energy: A Guide for the Perplexed*. Washington: Center for Strategic and International Studies and Peterson Institute for International Economics.

Smith, Anne E., Martin T. Ross, and W. David Montgomery. 2002. *Implications for Trading Implementation Design for Equity-Efficiency Trade-Offs in Carbon Permit Allocations*. Washington: Charles River Associates.

Stern, Todd, and William Antholis. 2007. "A Changing Climate: The Road Ahead for the United States." *Washington Quarterly*, Winter 2007–8, 175–88.

Stiglitz, Joseph E. 2006a. *Making Globalization Work*. New York: W. W. Norton.

———. 2006b. "A New Agenda for Global Warming." *Economists' Voice* 3, no. 7, article 3.

Van den Bossche, Peter. 2005. *The Law and Policy of the World Trade Organization*. Cambridge University Press.

World Bank. 2008. *International Trade and Climate Change: Economic, Legal, and Institutional Perspectives*. Washington.

World Resources Institute. 2008. "Climate Analysis Indicators Tool" (http://cait.wri.org [March 2009]).

JEFFREY A. FRANKEL 3

Addressing the Leakage/Competitiveness Issue in Climate Change Policy Proposals

Of all the daunting obstacles faced by the effort to combat global climate change, the problem of leakage is perhaps the most recent to gain serious attention from policymakers. Assume that a core of rich countries is able to agree for the remainder of the century on a path of targets for emissions of greenhouse gases (GHGs), following the lead of the Kyoto Protocol, or to agree on other measures to cut back on emissions, and that the path is aggressive enough at face value to go some way toward achieving the GHG concentration goals that environmental scientists say are necessary. Will global emissions in fact be reduced? Even under the business-as-usual scenario—that is, the path along which technical experts forecast that countries' emissions would increase in the absence of a climate change agreement—most emissions growth was expected to come from China and other developing countries. If these nations are not included in a system of binding commitments, global emissions will continue their rapid growth. But the problem is worse than that. Leakage means that emissions in the nonparticipating countries would actually rise above where they would otherwise be, thus working to undo the environmental benefits of the rich countries' measures. Furthermore, not wanting to lose "competitiveness" and pay economic costs for minor environmental benefits, the rich countries could lose heart and the entire effort could unravel. It is important to find ways to address concerns about competitiveness and leakage, but without undue damage to the world trading system.

The author would like to acknowledge useful input from Joe Aldy, Lael Brainard, Thomas Brewer, Steven Charnovitz, Juan Delgado, and Gary Sampson. He would further like to thank for support the Sustainability Science Program, funded by the Italian Ministry for Environment, Land and Sea, at the Center for International Development at Harvard University.

Developing Countries, Leakage, and Competitiveness

We need the developing countries inside the emissions control program, for several reasons.[1] As noted, these countries will be the source of the big increases in GHG emissions in coming years, according to the business-as-usual path. China, India, and other developing countries will represent up to two-thirds of global carbon dioxide emissions over the course of this century, vastly exceeding the expected contribution of countries belonging to the Organization for Economic Cooperation and Development of roughly one-quarter of global emissions. Without the participation of major developing countries, emissions abatement by industrial countries will not do much to mitigate global climate change.

If a quantitative international regime is implemented without the developing countries, their emissions are likely to rise even faster than the business-as-usual path, due to the problem of leakage. Leakage of emissions could come about through several channels. First, the output of energy-intensive industries could relocate from countries with emissions commitments to countries without. This could happen either if firms in these sectors relocate their plants to unregulated countries, or if firms in these sectors shrink in the regulated countries while their competitors in the unregulated countries expand. A particularly alarming danger is that a plant in a poor, unregulated country might use dirty technologies and thus emit more than a plant producing the same output would have in a high-standard, rich, regulated country, so that aggregate world emissions would actually go up rather than down!

Another channel of leakage runs via world energy prices. If participating countries succeed in cutting back their consumption of coal and oil, the high-carbon fossil fuels, demand will fall and the prices of these fuels will fall on world markets (other things equal). This is equally true if the initial policy is a carbon tax that raises the price to rich-country consumers as if it comes via other measures. Nonparticipating countries would naturally respond to declines in world oil and coal prices by increasing consumption.

Estimates vary regarding the damage in tons of increased GHG emissions from developing countries for every ton abated in an industrial country. But an

---

1. An additional reason we need developing countries inside the cap-and-trade system is to give the United States and other industrial countries the opportunity to buy relatively low-cost emissions abatement from developing countries, which is crucial to keep low the economic cost of achieving any given goal in terms of concentrations. This would increase the probability that industrial countries comply with the system of international emissions commitments. Elaboration is available from Aldy and Frankel (2004), Frankel (1998, 2005c, 2007), Seidman and Lewis (2009), and many other sources.

authoritative survey concludes "Leakage rates in the range 5 to 20 per cent are common."[2]

Even more salient politically than leakage is the related issue of competitiveness: American industries that are particularly intensive in energy or other GHG-generating activities will be at a competitive disadvantage to firms in the same industries operating in nonregulated countries.[3] Such sectors as aluminum, cement, glass, paper, chemicals, iron, and steel will point to real costs in terms of lost output, profits, and employment.[4] They will seek protection and are likely to get it.

The policy response to fears of leakage and competitiveness can take a variety of forms. *Tariffs* on imports of goods from producers who do not operate under emission regulations are perhaps the most straightforward, except that ascertaining carbon content is difficult. *Border tax adjustments* apply not just to import tariffs alone but to a combination of import tariffs and export subsidies. Broader phrases such as *trade controls, import penalties,* or *carbon-equalization measures* include the option—likely to be adopted in practice— of requiring importers to buy emission permits, or "international reserve allowances." For economists such *importer permit requirements* are precisely equivalent to import tariffs—the cost of the permit is the same as the tariff rate. Others would not so readily make this connection, however. International law may well defy economic logic by treating import tariffs as impermissible but permit requirements for imports as acceptable.[5] *Trade sanctions* go beyond trade controls: while the latter fall only on environmentally relevant sectors, the former target products that are arbitrary and unrelated to the noncompliant act, in an effort to induce compliance.[6]

2. Intergovernmental Panel on Climate Change (2001, chap. 8.3.2.3, pp. 536-44). In chapter 2 of this volume, Bordoff reports studies' estimates in the range of 8 to 11 percent, including an estimate from McKibbin and others (1999) that leakage if the United States adopted its Kyoto Protocol target unilaterally would have been 10 percent. Ho, Morgenstern, and Shih (2008) also find that the imposition of a price on carbon in the United States would produce substantial leakage for some industries, especially in the short run; they conclude that petrochemicals and cement are the most adversely impacted, followed by iron and steel, aluminum, and lime products. Demailly and Quirion (2008a) and Reinaud (2008) do not find large leakage effects from the first stage of the EU Emissions Trading System; but this tells us little about the next, much more serious, stage. I cannot help feeling that all these studies may underestimate some long-run general equilibrium effects.

3. It is not meaningful to talk about an adverse effect on the competitiveness of the American economy in the aggregate. Those sectors *low* in carbon intensity would in theory *benefit* from an increase in taxation of carbon relative to everything else. This theoretical point is admittedly not very intuitive. Far more likely to resonate publicly is the example that producers of renewable energy, and of the equipment that they use, would benefit from higher fossil fuel prices.

4. Houser and others (2008).

5. Pauwelyn (2007) and Fischer and Fox (2009).

6. They are used multilaterally only by the WTO and United Nations Security Council, and are not currently under consideration as a mechanism for addressing climate change (Charnovitz

The Possible Application of Trade Barriers by the United States

Of the twelve market-based climate change bills introduced in the 110th Congress, almost half called for some sort of border adjustments. Some would have featured a tax to be applied to fossil fuel imports. (This might be unobjectionable, *provided* the same tax is applied to the domestic production of the same fossil fuels; but otherwise it would be distortionary and illegal vis-à-vis the World Trade Organization.) Others would have required that energy-intensive imports surrender permits corresponding to the carbon emissions embodied in them.[7] The Bingaman-Specter Low Carbon Economy Act of 2007 would have provided that "if other countries are deemed to be making inadequate efforts [in reducing global GHG emissions], starting in 2020 the president could require importers from such countries to submit special emission allowances (from a separate reserve pool) to cover the carbon content of certain products." Similarly, the 2007 Lieberman-Warner Bill would have required the president to determine what countries have taken comparable action to limit GHG emissions; for imports of covered goods from covered countries, the importer would then have had to buy international reserve allowances.[8] In the 2007 bill, the requirement would have gone into effect in 2020. These requirements are equivalent to a tax on the covered imports. The two major presidential candidates in the 2008 U.S. election campaign supported some version of these bills, including import penalties in the name of safeguarding competitiveness vis-à-vis developing countries.

In addition, a different law that has already been passed and gone into effect poses similar issues: The Energy Independence and Security Act of 2007 "limits U.S. government procurement of alternative fuel to those from which the lifecycle greenhouse gas emissions are equal to or less than those from conventional fuel from conventional petroleum sources."[9] Canada's oil sands are vulnerable. Because Canada has ratified the Kyoto Protocol and the United States has not, the legality of this measure seems questionable, in the inexpert judgment of the author.

2003b, page 156). Pauwelyn (2008) compares some of these various options more carefully, from a legal standpoint. Fischer and Fox (2009) compare four of them from an economic standpoint: import tax alone, export rebate alone, full border adjustment, and domestic production rebate. Hufbauer, Charnovitz, and Kim (2009, chapter 3) are more exhaustive still.

7. Hufbauer, Charnovitz and Kim (2009, table 1.A.2).

8. The Lieberman-Warner Bill, S 2191: Americas Climate Security Act of 2007, sections 6005–6.

9. Energy Independence and Security Act of 2007, section 526; cited in *Financial Times*, March 10, 2008. Pauwelyn (2008) deals with government procurement and Kyoto.

The Possible Application of Trade Barriers by the EU

It is possible that many in Washington do not realize that the United States is likely to be the victim of legal sanctions before it is the wielder of them. In Europe, firms have already entered the first Kyoto budget period of binding emission limits, competitiveness concerns are well advanced, and the nonparticipating United States is an obvious target of resentment.[10]

After the United States failed to ratify Kyoto, European parliamentarians in 2005, and French prime minister Dominique de Villepin in 2006, proposed a "Kyoto carbon tax" or "green tax" against imports from the United States.[11] The European Commission had to make a decision on the issue in January 2008, when the European Union determined its emission targets for the post-Kyoto period. In preparation for this decision, French president Nicolas Sarkozy warned:

> If large economies of the world do not engage in binding commitments to reduce emissions, European industry will have incentives to relocate to such countries. . . . The introduction of a parallel mechanism for border compensation against imports from countries that refuse to commit to binding reductions therefore appears essential, whether in the form of a tax adjustment or an obligation to buy permits by importers. This mechanism is in any case necessary in order to induce those countries to agree on such a commitment.[12]

The mechanism envisioned here sounds similar to that in the Bingaman-Specter and Lieberman-Warner bills, with the difference that it could go into effect soon, because Europe is already limiting emissions whereas the United States is not. In the event, the EU Commission instead included this provision in its directive:

> Energy-intensive industries which are determined to be exposed to significant risk of carbon leakage could receive a higher amount of free allocation or an effective carbon equalization system could be introduced with a view to putting EU and non-EU producers on a comparable footing. Such a system could apply to importers of goods requirements similar to those applicable to installations within the EU, by requiring the surrender of allowances.[13]

---

10. Bierman and Brohm (2005); Bhagwati and Mavroidis (2007); National Board of Trade, Government of Sweden (2004). Recent papers that compare the various options for border measures in a European context include Alexeeva-Talebi, Loschel and Mennel (2008), Demailly and Quiron (2009), and Reinaud (2008).

11. Beattie (2008); "Mandelson Rejects CO2 Border Tax," EurActiv.com, December 18, 2006.

12. Letter to EU Commission president José Manuel Barroso, January 2008.

13. The source for this is paragraph 13 of the "Directive of the European Parliament and of the Council Amending Directive 2003/87/EC so as to Improve and Extend the EU Greenhouse Gas Emissions Allowance Trading System," January 2008.

The second of the two options, "carbon equalization," sounds consistent with what is appropriate—and with the sort of measures suggested by Sarkozy and spelled out in detail in the U.S. bills. The first option is poorly designed, however.

Free allocation of permits would help European industries that are carbon intensive and therefore vulnerable to competition from nonmembers by giving them a larger quantity of free GHG emission permits. According to simple microeconomic theory, this would do nothing to address leakage. Because carbon-intensive production is cheaper in nonparticipating countries, the European firms in theory would simply sell the permits they receive and pocket the money, with the carbon-intensive production still moving from Europe to the nonparticipants. Admittedly, in practice there might be some effects; for example, an infusion of liquidity might keep in operation a firm that otherwise would go bankrupt. But overall, there would probably be almost as much leakage as if there had been no policy response at all.[14] Presumably, the purpose behind this option is not to minimize leakage, for which it would be the wrong remedy, nor even to punish nonparticipating countries, but simply to buy off domestic interests so that they will not oppose action on climate change politically.

Would Trade Controls or Sanctions Be Compatible with the WTO?

Would measures that are directed against carbon dioxide emissions in other countries, as embodied in electricity or in goods produced with it, be acceptable under international law? Not many years ago, most international experts would have said that import barriers against carbon-intensive goods, whether tariffs or quantitative restrictions, would necessarily violate international agreements. Under the General Agreement on Tariffs and Trade (GATT), although countries could use import barriers to protect themselves against environmental damage that would otherwise occur within their own borders, they could not use import barriers in efforts to affect how goods are produced in foreign countries, so-called process and production methods (PPMs). A notorious example was the GATT ruling against U.S. barriers to imports of tuna from dolphin-unfriendly Mexican fishermen. But things have changed.

The World Trade Organization (WTO) came into existence, succeeding the GATT, at roughly the same time as the Kyoto Protocol. The drafters of each treaty showed more consideration for the other than do the rank and file among environmentalists and free traders, respectively. The WTO regime is more respectful of the environment than was its predecessor. Article XX allows

---

14. Of course, free allocation of permits would be an equally bad way of protecting exposed industries in the United States. See the discussion by Bordoff in chapter 2 in this volume.

exceptions to Articles I and III for purposes of health and conservation. The Preamble to the 1995 Marrakech Agreement establishing the WTO seeks "to protect and preserve the environment;" and the 2001 Doha Communiqué that sought to start a new round of negotiations declares: "The aims of . . . open and non-discriminatory trading system, and acting for the protection of the environment . . . must be mutually supportive." The Kyoto Protocol text is equally solicitous of the trade regime. It says that the parties should "strive to implement policies and measures . . . to minimize adverse effects . . . on international trade." The United Nations Framework Convention on Climate Change features similar language.

GHG emissions are PPMs. Is this an obstacle to the application measures against them at the border? I do not see why it has to be. Three precedents can be cited: sea turtles in the Indian Ocean, ozone in the stratosphere, and tires in Brazil.

The true import of a 1998 WTO panel decision on the shrimp-turtle case was missed by almost everyone. The big significance was a pathbreaking ruling that environmental measures can target not only exported products (Article XX) but also partners' PPMs—subject, as always, to nondiscrimination (Articles I and III). The United States was in the end able to seek to protect turtles in the Indian Ocean, provided it did so without discrimination against Asian fishermen. Environmentalists failed to notice or consolidate the PPM precedent, and to the contrary were misguidedly up in arms over this case.[15]

Another important precedent was the Montreal Protocol on stratospheric ozone depletion, which contained trade controls. The controls had two motivations.[16] The first was to encourage countries to join. And the second, if major countries had remained outside, was to minimize leakage, the migration of the production of banned substances to nonparticipating countries. In the event, the first worked, so the second was not needed.

In case there is any doubt that Article XX, which uses the phrase "health and conservation," also applies to global environmental concerns such as climate change, a third precedent is relevant. In 2007, a new WTO Appellate Body decision regarding Brazilian restrictions on imports of retreaded tires confirmed the applicability of Article XX(b): Rulings "accord considerable flexibility to WTO Member governments when they take trade-restrictive measures to protect life or health . . . [and] apply equally to issues related to trade and environmental protection, . . . including measures taken to combat global warming."[17]

---

15. For a full explanation of the legal issues, see Charnovitz (2003a). Also see Bhagwati and Mavroidis (2007), Charnovitz and Weinstein (2001), Deal (2008), and Weinstein (2001).
16. Brack (1996).
17. From a personal communication with Brendan McGivern, December 12, 2007.

These three examples go a long way toward establishing the legitimacy of trade measures against PPMs. Many trade experts, both economists and international lawyers, are not yet convinced[18]—let alone representatives of India and other developing countries. I personally have come to believe that the Kyoto Protocol could have followed the Montreal Protocol by incorporating well-designed trade controls aimed at nonparticipants. One aspect that strengthens the applicability of the precedent is that we are not talking about targeting practices in other countries that harm solely the local environment, where the country can make the case that this is nobody else's business. The depletion of the stratospheric ozone and the endangerment of sea turtles are global externalities. (It helped that these are turtles that migrate globally.) So is climate change from GHG emissions. A ton of carbon emitted into the atmosphere hurts all residents of the planet.

Principles for Designing Legitimate Penalties on Carbon-Intensive Imports

Although the shrimp-turtle case and the Montreal Protocol help establish the principle that well-designed trade measures can legitimately target PPMs, at the same time they suggest principles that should help guide drafters as to what is good design. First, the existence of a multilaterally negotiated international treaty such as the Kyoto Protocol conditions the legitimacy of trade controls. On the one hand, that leakage to nonmembers could negate the goal of the protocol strengthens the case for (the right sort of) trade measures. It is stronger, for example, than in the shrimp-turtle case, which was largely a unilateral U.S. measure.[19] On the other hand, the case for trade measures is weaker than it was for the Montreal Protocol. (Multilateral initiatives like the latter are on firmer ground than unilateral initiatives.) The Kyoto Protocol could have made explicit allowance for multilateral trade controls, and chose not to. The case would be especially weak for American measures if the United States has still not ratified the protocol or a successor agreement. The Europeans have a relatively good case against the United States, until such time as the United States ratifies. But the case would be stronger still if a future multilateral agreement—for example, under the United Nations Framework Convention on

18. Some experts believe that even multilateral trade penalties against nonmembers might not be permissible under the WTO. See Sampson (2000, 87).

19. Webster (2008) explains that unilateral measures more likely acceptable if in pursuit of an existing multilateral agreement such as the Kyoto Protocol. Even sea turtles are, however, given some protective status by their inclusion in Appendix 1 of the Convention on International Trade in Endangered Species of Wild Fauna and Flora.

Climate Change (UNFCCC)—agreed on the legitimacy of trade measures and on guidelines for their design.

Second, there is the question of the sorts of goods or services to be made subject to penalty. It would certainly be legitimate to apply tariffs against coal itself, assuming that the domestic taxation of coal or a domestic system of tradable permits was in place. It is probably also legitimate when applied to the carbon content of electricity, though this requires acceptance of the PPM principle. The big question is the carbon/energy content of manufactures. Trade sanctions would probably not be legitimate when applied solely as punishment for free riding, against unrelated products of a nonmember, or, in a more extreme case, on clean inputs—for example, a ban on U.S. turbines used for low-carbon projects (unless perhaps economy-wide sanctions were multilaterally agreed by UNFCCC members).[20]

Paradoxically, the need to keep out coal-generated electricity or aluminum from nonmembers of the Kyoto Protocol is greater than the need to keep out coal itself. The reason is that the Protocol already puts limits on within-country emissions. Assuming the limits are enforced, then the world community has no particular interest in how the country goes about cutting its emissions. But if the country imports coal-generated electricity or aluminum from nonmembers, the emissions occur outside its borders and the environmental objective is undermined.

It is hard to determine carbon content of manufactures. In practice, the best would be to stay with the half-dozen biggest-scale, most-energy-intensive industries—probably aluminum, cement, steel, paper, glass, perhaps iron and chemicals. Even here there are difficult questions. What if the energy used to smelt aluminum in another country is cleaner than in the importing country (Iceland's energy comes from hydropower and geothermal power) or dirtier (much of Australia's energy comes from coal)? How can one distinguish the marginal carbon content of the energy used for a particular aluminum shipment from the average carbon content of energy in the country of origin? These are questions that will have to be answered. But as soon as one goes beyond six or seven big industries, it becomes too difficult for even a good-faith investigator to discern the effective carbon content, and the process is also too liable to abuse. One would not want to levy tariffs against the car parts that are made with the metal that was produced in a carbon-intensive way, or against the auto-

---

20. Charnovitz (2003b, 156) emphasizes the distinction between trade controls, which fall on environmentally relevant sectors, versus trade sanctions, where the targeted products are arbitrary and unrelated to the noncompliant act (and are used multilaterally only by the WTO and UN Security Council).

mobiles that used those car parts (they could be high-mileage hybrids!), or against the products of the firms that bought the cars, and so on.[21]

The Big Danger

Just because a government measure is given an environmental label does not necessarily mean that it is motivated primarily—or even at all—by bona fide environmental objectives. To see the point, one has only to look at the massive mistake of American subsidies of ethanol made from corn (and protection against competing sugar imports from Brazil). If each country on its own imposes border penalties on imports in whatever way suits national politics, they will be poorly targeted, discriminatory, and often disguisedly protectionist. When reading the language in the U.S. congressional bills or the EU decision, it is not hard to imagine that special interests could take over for protectionist purposes the process whereby each government decides whether other countries are doing their share or what foreign competitors merit penalties.[22] Thus the competitiveness provisions will indeed run afoul of the WTO, and they will deserve to.

It is important who makes the determinations regarding what countries are abiding by carbon-reduction commitments, what entity can retaliate against the noncompliers, what sectors are fair game, what sort of barriers are appropriate, and when a target country has moved into compliance so that it is time to remove the penalty. One policy conclusion is that these decisions should be delegated to independent panels of experts rather than made by politicians.

The most important policy conclusion is that we need a multilateral regime to guide such measures. Ideally, such a regime would be negotiated along with a successor to the Kyoto Protocol that set targets for future periods and brought the United States and developing countries inside. But if that negotiation process takes too long, it might be useful in the shorter run for the United States to enter negotiations with the European Union to harmonize guidelines for border penalties, ideally in informal association with the secretariats of the UNFCCC and the WTO.[23]

---

21. The 2008 revision of the McCain-Lieberman Bill broadened "covered products" to include goods that generate emissions more indirectly (see chapter 2 in this volume).

22. The congressional language imposing penalties on imports from countries that do not tax carbon was apparently influenced by the International Brotherhood of Electrical Workers, which regularly lobbies for protection of American workers from foreign competition; see Beattie (2008). Simultaneously, the European Trade Union Confederation urged the EU Commission to tax imports from countries refusing to reduce emissions. See *Wall Street Journal* (2008).

23. Sampson (1999).

Why Take Multilateralism Seriously?

"Why should WTO obligations be taken seriously?" some may ask. There are three possible answers, based on considerations of international citizenship, good policy, and realpolitik.

First, with regard to international citizenship, one question is whether the United States wants to continue the drift of the recent past in the general direction of international rogue country status or rather return to the highly successful postwar strategy of adherence to international law and full membership in—indeed, leadership of—multilateral institutions. The latter course does not mean routinely subordinating U.S. law, let alone American interests, to international law. There will be cases where the United States wants to go its own way. But the effort on climate change should surely not be one of these cases. Among other reasons is the fact that GHG emissions are inherently a global externality. No single country can address the problem on its own, due to the free rider problem. Although there is a role for unilateral actions on climate change—for example, by the United States, as part of a short-term effort to demonstrate seriousness of purpose and begin to catch up with the record of the Europeans—in the long term, multilateral action offers the only hope of addressing the problem. The multilateral institutions are already in place—specifically, the UNFCCC; its child, the Kyoto Protocol; and the WTO—and they were predominantly created under U.S. leadership.

Second, the basic designs and operations of these institutions happen to be relatively sensible, taking political realities as given. They are more sensible than most critics of the international institutions and their alleged violations of national sovereignty believe. This applies whether the critics are on the left or right, and whether their main concern is the environment or the economy.[24] One can place very heavy weight on economic goals and yet realize the desirability of addressing externalities, minimizing leakage, dealing with competitiveness concerns, and so forth. One can place very heavy weight on environmental goals and yet realize the virtues of market mechanisms, nondiscrimination, reciprocity, addressing international externalities *cooperatively*, preventing special interests from hijacking environmental language for their own financial gain, and so forth.

The third reason why the United States should be prepared to modify the sort of "international reserve allowances" language of the Lieberman-Warner Bill, and to move in the direction of multilateral coordination of guidelines for

24. I have addressed elsewhere other ways in which the climate regime (Kyoto) could come into conflict with the trade regime (WTO), and the more general questions of whether free trade and environmental protection need be in conflict—Frankel (2004, 2005a, 2005c).

such measures, comes from hardheaded self-interest: a desire to avoid being the victim of emulation or retaliation. Section 6006 of Lieberman-Warner did not envision these measures going into effect until 2020.[25] This is as it should be, because any such bill must give the United States time to start playing the game before it can presume to punish other players for infractions. But the EU language could be translated into penalties against U.S. products any day. The EU members are far from the only governments that could claim to have taken stronger climate change policies than the United States.[26] It is in the American interest to have any border penalties governed by a sensible system of multilateral guidelines. The European Union might welcome U.S. participation in joint negotiations to agree on guidelines, as part of a process of negotiations over the Kyoto successor regime.

The argument is stronger than the historical examples of U.S. import barriers leading to subsequent emulation and retaliation that come back to hit American exports (the Smoot Hawley tariff in 1930, antidumping cases in the 1980s, . . .). Here the United States has an opportunity to influence others' barriers against its goods ten years before it would be putting up barriers against theirs.

Concluding Recommendations

Both the economics and the law are complicated. The issues need further study. Nevertheless, this chapter offers a central message: Border measures to address leakage need not necessarily violate WTO law or sensible trade principles, but there is a very great danger in practice that they will.

I conclude with several subjective judgments as to principles that could guide a country's border measures if its goal were indeed to reduce leakage and to avoid artificially tilting the playing field toward carbon-intensive imports from nonparticipating countries or damaging the world trading system, especially if it is viewed as politically necessary to do something to address competitiveness concerns. I classify characteristics of possible border measures into three categories, which I will name by color (for lack of better labels):

—The "white" category: those that seem to me reasonable and appropriate.[27]

25. The Boxer-Lieberman-Warner substitute version (S 3036), voted on in June 2008, moved the starting date for border adjustments forward to 2014.

26. Even China has apparently enacted efficiency standards on automobiles, refrigerators, and air conditioners that exceed regulations in the United States. How will Americans react if China puts justified penalties on imports from the United States?

27. Hufbauer, Charnovitz, and Kim (2009, chapter 5) call this category "the green space" and present a list of desirable attributes which is more authoritative than the one I had drawn up, at

—The "black" category: those that seem to me very dangerous, in that they are likely to become an excuse for protectionism.

—The "gray" category: those that fall in between.

The white (appropriate) border measures could be either tariffs or (equivalently) a requirement for importers to surrender tradable permits. These principles include:

—Measures should follow some multilaterally agreed-to set of guidelines among countries participating in the emission targets of the Kyoto Protocol and/or its successors.

—Judgments as to findings of fact—which countries are complying or not, which industries are involved and what is their carbon content, which countries are entitled to respond with border measures, and the nature of the response—should be made by independent panels of experts.

—Measures should be applied only by countries that are reducing their emissions in line with the Kyoto Protocol and/or its successors, against countries that are not, either due to refusal to join or failure to comply.

—Border tax adjustments should target only imported fossil fuels, and a half dozen of the most energy-intensive major import-competing industries: aluminum, cement, steel, paper, and glass, and perhaps iron and chemicals.

The black (inappropriate) border measures include:

—Unilateral measures applied by countries that are not participating in the Kyoto Protocol or its successors.

—Calculations of carbon content of imports by formulas that presume firms necessarily use the same production processes as domestic competitors.[28]

—Judgments as to findings of fact that are made by politicians, who are vulnerable to political pressure from interest groups for special protection.

—Unilateral measures that seek to sanction an entire country, rather than targeting narrowly defined energy-intensive sectors.

—Import barriers against products that are further removed from the carbon-intensive activity, such as firms that use inputs that are produced in an energy-intensive process.

---

least from a legal standpoint. Green is the more familiar color, but I had thought to avoid it because of possible confusion with the "green box" of the WTO's Agreement on Agriculture.

28. In the Venezuelan reformulated gasoline case, the WTO panel ruled that the United States should have allowed for differences in foreign firms' production processes. Venezuela successfully claimed that U.S. law violated national treatment—that is, discriminated in favor of domestic producers with regard to whether refineries were allowed to use individual composition baselines when measuring pollution reduction. Pauwelyn (2007) proposes that if the foreign producer does not voluntarily provide the information needed to calculate carbon content, then as a back-up the U.S. Customs Bureau assign imports an implicit carbon content based on the production techniques that are dominant in the United States.

—Subsidies—whether in the form of money or extra permit allocations—to domestic sectors that are considered to have been put at a competitive disadvantage.

The gray (intermediate) measures include:

—Unilateral measures that are applied in the interim before there has been time for multilateral negotiation over a set of guidelines for border measures.

—The import penalties might follow the form of existing legislation on countervailing duties.

Comment

Comment by Joseph E. Aldy

In chapter 3 of this volume, Jeffrey Frankel thoughtfully addresses the question of how to design effective policies to mitigate the adverse competitiveness effects that result from domestic climate change policy. In evaluating measures to address the climate-competitiveness issue, three categories of questions merit consideration. First, how serious are the competitiveness effects? Do they justify policy intervention? Second, are the policy responses to competitiveness effects that are being considered adequate to remedy these effects? Do they effectively level the playing field? Third, what are the implications of such policy responses to competitiveness effects for international negotiations? Do they promote broader participation or risk a trade war?

Before addressing these questions, let us be clear by what is meant by the "competitiveness effects of climate change policy." Aldy and Pizer define such competitiveness effects as the "adverse business impacts related to a domestic GHG regulation *and* the absence of regulation on international competitors."[1] In economic terms, such competitiveness effects reflect either the relocation of firms and manufacturing facilities to countries without domestic climate change policy or the allocation of investment in new facilities away from countries with climate policies to those countries without such policies. In environmental terms, this reflects a reduction in environmental benefits as greenhouse gas (GHG) emissions increase in countries without climate policies, partially offsetting the emission abatement in countries with climate change policies. Some have described this as "America's China problem"—if the United States moves forward with domestic climate change policy while China does not. Of course, this is not a uniquely American phenomenon. The French refer to this as their American problem (as alluded to by Frankel), because France and the European Union have moved forward with the EU's Emission Trading Scheme

1. Aldy and Pizer (2008, 24).

(ETS) and other efforts under the Kyoto Protocol, an agreement from which the United States walked away.

How serious are these competitiveness effects? Frankel refers to the Intergovernmental Panel on Climate Change's assessment of the literature, which shows GHG emission leakage could be on the order of 5 to 20 percent.[2] In the aggregate, competitiveness effects may not be that serious. Approximately 70 percent of U.S. carbon dioxide emissions occur in nontradable sectors, such as transportation and residential and commercial buildings.[3] It seems unlikely that imposing a price on carbon would cause a migration of Americans to developing countries. The likely sources of near-term, low-cost emission abatement also occur primarily in nontradable sectors. Only about 15 percent of the estimated emission abatement in 2015 would likely occur outside of the nontradable transportation and buildings sectors.[4] Even within the manufacturing sector, firm relocation in response to environmental regulatory costs will be less likely for those producing goods with high transport costs or that have a large physical structure share of their capital.[5]

The climate-competitiveness effects could accrue primarily to the most-energy-intensive firms facing international competition. We have recently investigated this issue by simulating the effects of a unilateral carbon pricing policy on the manufacturing sector in 2015.[6] We undertook an econometric analysis of the historic relationship between energy prices and various measures of competitiveness, such as employment, production, and consumption (defined as production plus net imports) for the more than four hundred manufacturing industries in the United States. The analysis is careful in controlling for a variety of factors that could affect industrial competitiveness, and explicitly accounts for the energy intensity of each industry's manufacturing in estimating the energy price-competitiveness relationship. Using these estimated relationships, we simulated the effects of a $15 per ton carbon dioxide (CO_2) price.

Table 3C-1 presents the summary results for our analyses, aggregated to several of the most-well-known, energy-intensive industry groups. A $15 per ton CO_2 price would increase the price of electricity in the manufacturing sector about 8 percent in 2015.[7] We estimate that overall manufacturing employment

2. As Frankel appropriately notes, this could reflect firm/facility relocation (competitiveness effects) as well as the effects of declining prices in world energy markets as developed countries pursue mitigation policies. Developing countries could respond to these lower energy prices and increase their emissions above what they would have been otherwise.

3. U.S. Energy Information Administration (2007).

4. U.S. Energy Information Administration (2006). This estimate also reflects the emission abatement in the utility sector, but attributed by share of consumption to end-use sectors.

5. Ederington, Minier, and Levinson (2005).

6. Aldy and Pizer (2009).

7. U.S. Energy Information Administration (2008).

Table 3C-1. Predicted Effects of a \$15 per Ton Carbon Dioxide Price on Various Manufacturing Sectors

Percent

Industry	Employment	Production	Consumption	Competitiveness
Industrial chemicals	−1.5	−2.7	−1.8	−0.9
Paper	−2.1	−3.3	−2.4	−0.9
Iron and steel	−1.6	−2.7	−1.9	−0.8
Aluminum	−1.0	−2.0	−1.4	−0.7
Cement	−0.4	−1.6	−0.9	−0.7
Bulk glass	−2.3	−3.4	−2.7	−0.6
Manufacturing average	−0.2	−1.3	−0.6	−0.7

Source: Aldy and Pizer (2009).

Note: Consumption is defined as domestic production plus net imports. The competitiveness effect is defined as the difference between production and consumption effects.

would decline on average about 0.2 percent, although it is important to note that the 80 percent most-energy-efficient industries do not experience a change that we would statistically discern from zero. The more energy-intensive industries, however, experience larger declines in employment up to 2.3 percent in bulk glass. The effects on production are more pronounced than they are for employment, but they follow similar trends between total manufacturing average effects and energy-intensive manufacturing, as well as among those energy-intensive industries.

Although production in the most-energy-intensive industries declines between about 1.6 to 3.4 percent under a \$15 per ton CO_2 price, competition through international trade appears to constitute a modest fraction of this impact. Domestic consumption of goods produced in the energy-intensive industries also declines with this carbon price, on the order of 0.9 to 2.7 percent. Our analysis shows that consumers of goods produced in energy-intensive industries would not switch to imports under a domestic climate policy; they would reduce their consumption of those goods. The net effect of international competition, represented by the rightmost column in the table, is about 0.6 to 0.9 percentage points of the 1.6 to 3.4 percentage decline in production. It is not so much "us versus them" as it is "us versus us." In effect, the consumers of energy-intensive goods respond to the higher cost (because of the carbon price) by becoming more efficient in how they use these goods, and they substitute to less energy-intensive alternatives, not by importing the same goods from foreign firms.

This analysis suggests that the competitiveness effect of climate policy could be relatively modest, even on the most-energy-intensive industries. Despite the fact that this analysis raises questions about the need for policy responses to climate-competitiveness pressures, the debate about the nature of such policy responses is likely to continue. For such policy responses to adequately elim-inate competitiveness effects, they must eliminate the CO_2 price wedge between U.S. firms and their foreign competitors. This price wedge effectively drives the relocation of economic activity from higher carbon price regions to zero- or low-carbon price regions. Several proposals to address competitiveness con-cerns have been tabled in the U.S. policy debate, but none adequately eliminates the carbon price wedge.

The Lieberman-Warner Bill includes a variant of the proposal from the Inter-national Brotherhood of Electrical Workers and American Electric Power that would require GHG emission allowances to cover the embedded carbon of bulk commodity imports if they originate in countries found to have domestic cli-mate change policies deemed to be insufficiently comparable to the U.S. cap-and-trade program.[8] Firms importing these goods could buy emission allowances from the U.S. government, or possibly from certain recognized cap-and-trade programs (for example, the ETS) or emissions reduction credit programs (for example, the Kyoto Protocol's Clean Development Mechanism). The U.S. government would supply emission allowances from a separate fund than the allowances constituting the U.S. economy-wide cap under the bill, but the import emission allowances would be priced at the same level as the market-clearing price in the U.S. emission trading market. In effect, this is a border tax by another name. If these imports had to hold allowances to cover every ton of embedded carbon, then this would close the competitiveness wedge. The pro-posal, however, recognizes that if energy-intensive firms in the United States receive at least a partially gratis allocation (which they do under Lieberman-Warner and other bills), then so should importers in order to satisfy World Trade Organization rules. For example, suppose that U.S. firms receive 20 percent of the program's allowances for free and consider an importer trying to bring in a rolled steel shipment with 1,000 tons of embedded carbon. The importer only needs to buy allowances to cover 800 tons of carbon, not the full 1,000 tons, because of the need to treat foreign and domestic firms equally under the World Trade Organization. The effective carbon price on this shipment then is only 80 percent of the price domestic firms face. If the importer decides to buy allowances from the EU's ETS or the Clean Development Mechanism, which it would only do if those prices were lower than the going price in the U.S. market, then, again, the importer would face a lower price than what domestic firms face.

---

8. This discussion will focus on the proposal presented in the 2008 Manager's Amendment to the Lieberman-Warner Bill (S 3036), not the 2007 Lieberman-Warner Bill version (S 2191) or the Bingaman-Specter Bill (S 1766) version of this proposal.

The prospect of a gratis allowance allocation to compensate energy-intensive firms has also received some traction in the policy debate. Free allowances to firms will not affect the carbon price wedge. As the EU's ETS showed, even with a free allowance allocation, energy prices increase with the emission trading market allowance price. The carbon price embedded in goods is effectively the same regardless of whether allowances are given away or auctioned.[9] The free allocation will affect the profits of the recipient firms; a free allowance allocation is simply a transfer of an asset from the government to the private sector. Thus, freely giving away an asset will affect firms' market entry and exit decisions. This may be sufficient, from a political economy perspective, to secure the support of energy-intensive firms, but it is insufficient to eliminate the competitiveness effects.[10]

The steel industry has advocated for performance standards on imports.[11] To bring imports into the U.S. economy, foreign firms would have to meet a carbon-intensity standard for the production of that class of goods. Such an approach returns environmental policy to the early 1970s, when cost-effectiveness played virtually no role in policy design. It risks increasing substantially the costs to users of energy-intensive imports, and it would increase the total costs for a given amount of GHG emission abatement. Finally, it is quite unlikely that such a standards approach would eliminate the competitiveness price wedge because it works through a mechanism unrelated to carbon prices. In all three cases, the measures intended to address competitiveness do not effectively level the playing field.

Designing such competitiveness policy measures in a domestic climate change policy regime will have implications for international negotiations. Some countries may react quite negatively to unilateral U.S. action on this front. For example, the European Union could simply adopt the same competitiveness policy the United States has designed to address its concern with developing countries but apply it to America during the Kyoto Protocol commitment period of 2008–12 (see chapter 3 for an elaboration of this point). The Lieberman-Warner Bill debated in June 2008 gave developing countries all of two years after the start of the U.S. cap-and-trade program to develop comparable domestic policies before the import GHG emission allowance requirement would become operative. Because the United States failed to ratify the Kyoto Protocol and has stayed on the sidelines in terms of mandatory domestic emission mitigation efforts, some countries could consider such an approach as con-

9. See Aldy and Pizer (2008) for more discussion of this issue, and the minor exceptions to this rule.
10. Frankel takes a strong exception to such subsidies through free allowance allocations.
11. Slater (2008).

frontational from a (developed country) latecomer to the climate change mitigation effort. Though some believe that this threat could bring large developing countries to the negotiating table, others fear that it could cause China, India, and other large countries to walk away from the climate talks and consider trade retaliations that could descend into a trade war. Or perhaps China might simply find clever ways to circumvent the burdens of the U.S. policy. Suppose South Korea implemented a domestic climate change policy deemed to be comparable to the U.S. effort. A ship in Shanghai could load up with Chinese steel, sail to Busan, and unload its Chinese steel. Then the ship could load up with Korean steel and take it to the U.S. market. It may be unlikely to expect, however, that China would impose domestic regulatory burdens on its entire steel sector to protect the small fraction that exports to the United States.

The risks posed to the international climate and international trade regimes by a unilateral approach to addressing climate-competitiveness concerns suggest the need for a more thoughtful, multilateral consideration of these issues. The Montreal Protocol, which has successfully resulted in the phasing out of many ozone-depleting substances, includes trade sanctions to enforce compliance. Though the strong record of compliance with the Montreal Protocol's goals has obviated the need to use the sanctions, this multilateral approach may be a better avenue for addressing competitiveness under climate change policy. Frankel's recommendation for a multilateral approach under the United Nations Framework Convention on Climate Change is quite sound and merits serious consideration by U.S. policymakers and negotiators around the world.

References

Aldy, Joseph, and Jeffrey Frankel. 2004. "Designing a Regime of Emission Commitments for Developing Countries That Is Cost-Effective and Equitable." Paper written for conference on G20 Leaders and Climate Change, Council on Foreign Relations, New York, September 20–21.

Aldy, J. E., and W. A. Pizer. 2008. *Issues in the Design of U.S. Climate Change Policy.* RFF Discussion Paper 08-20. Washington: Resources for the Future.

Aldy, J.E., and W.A. Pizer. 2009. *The Competitiveness Impacts of Climate Change Mitigation Policies.* Arlington, VA: Pew Center on Global Climate Change, May.

Aldy, Joseph, Scott Barrett, and Robert Stavins. 2003. "Thirteen Plus One: A Comparison of Global Climate Architectures." *Climate Policy* 3, no. 4: 373–97.

Alexeeva-Talebi, Victoria, Andreas Loschel, and Tim Mennel. 2008. "Climate Policy and the Problem of Competitiveness: Border Tax Adjustments or Integrated Emissions Trading?" Discussion Paper 08-061, Zentrum fur Europaische Wirtschaftsforschung GmbH, Mannheim, Germany.

Beattie, Alan. 2008. "Green Barricade: Trade Faces a New Test as Carbon Taxes Go Global." *Financial Times*, January 24.

Bhagwati, Jagdish, and Petros C. Mavroidis. 2007. "Is Action Against U.S. Exports for Failure to Sign the Kyoto Protocol WTO Legal?" *World Trade Review* 6: 299–310.

Bierman, Frank, and Rainer Brohm. 2005. "Implementing the Kyoto Protocol without the United States: The Strategic Role of Energy Tax Adjustments at the Border." *Climate Policy* 4, no. 3: 289–302 .

Brack, D. 1996. *International Trade and the Montreal Protocol.* London: Royal Institute of International Affairs and Earthscan Publications.

Charnovitz, Steven. 2003a. "The Law of Environmental 'PPMs' in the WTO: Debunking the Myth of Illegality." *Yale Journal of International Law* 27, no. 1: 59–110.

———. 2003b. "Trade and Climate: Potential Conflicts and Synergies." In *Beyond Kyoto: Advancing the International Effort against Climate Change.* Washington: Pew Center on Global Climate Change.

Charnovitz, Steven, and Michael Weinstein. 2001. "The Greening of the WTO." *Foreign Affairs* 80, no. 6: 147–56.

Cosbey, Aaron, Sergio Saba, and Lucas Assuncao. 2003. "Options on Kyoto under WTO Rules." In *Implications, Including for Development, of the Interface between Environment and Trade Policies for Oil-Exporting Countries.* New York: United Nations.

Deal, Timothy. 2008. *WTO Rules and Procedures and Their Implication for the Kyoto Protocol.* New York: United States Council for International Business.

Demailly, D., and P. Quirion. 2008a. "European Emission Trading Schemes and Competitiveness: A Case Study on the Iron and Steel Industry." *Energy Economics* 30: 2009-2027.

———. 2009. "Leakage from Climate Polices and Border Tax Adjustment: Lessons from a Geographic Model of the Cement Industry." In *The Design of Climate Policy,* ed. R. Guesnerie and H. Tulkens. MIT Press.

Ederington, J., J. Minier, and A. Levinson. 2005. "Footloose and Pollution-Free." *Review of Economics and Statistics* 87, no. 1: 92–99.

Fischer, Carolyn, and Alan Fox. 2009. "Comparing Policies to Combat Emissions Leakage: Border Tax Adjustment versus Rebates." RFF Discussion Paper 09-02. Washington: Resources for the Future.

Frankel, Jeffrey. 1998. "The Kyoto Agreement on Global Climate Change: The Administration Economic Analysis, Remarks to the NBER Conference on Tax Policy and the Economy," Oct. 20 (http://ksghome.harvard.edu/~jfrankel/speechs&testimony.htm [May 2009]).

———. 2004. "Kyoto and Geneva: Linkage of the Climate Change Regime and the Trade Regime." Paper presented at Broadening Climate Discussion: The Linkage of Climate Change to Other Policy Areas, conference sponsored by Fondazione Eni Enrico Mattei and Massachusetts Institute of Technology, Venice, June.

———. 2005a. "Climate and Trade: Links between the Kyoto Protocol and WTO." *Environment* 47, no. 7: 8–19.

———. 2005b. "The Environment and Globalization." In *Globalization: What's New*, ed. Michael Weinstein. Columbia University Press.

———. 2005c. "You're Getting Warmer: The Most Feasible Path for Addressing Global Climate Change Does Run Through Kyoto." In *Trade and Environment: Theory and Policy in the Context of EU Enlargement and Transition Economies*, ed. John Maxwell and Rafael Reuveny. Cheltenham, U.K.: Edward Elgar.

———. 2007. "Formulas for Quantitative Emission Targets." In *Architectures for Agreement: Addressing Global Climate Change in the Post Kyoto World*, ed. Joseph E. Aldy and Robert Stavins. Cambridge University Press.

Frankel, Jeffrey, and Andrew Rose. 2005. "Is Trade Good or Bad for the Environment? Sorting out the Causality." *Review of Economics and Statistics* 87, no. 1: 85–91.

Ho, Mun, Richard Morgenstern, and Jhih-Shyang Shih. 2008. "Impact of Carbon Price Policies on U.S. Industry." RFF Discussion Paper 09-37. Washington: Resources for the Future.

Houser, Trevor, Rob Bradley, Britt Childs, Jacob Werksman, and Robert Heilmayr. 2008. *Leveling the Carbon Playing Field: International Competition and US Climate Policy Design*. Washington: Peterson Institute for International Economics.

Hufbauer, Gary, Steve Charnovitz, and Jisun Kim. 2009. *Global Warming and the World Trading System*. Washington: Peterson Institute for International Economics.

Intergovernmental Panel on Climate Change. 2001. *Third Assessment Report: Climate Change 2001*, ed. Working Group III. Geneva.

McKibbin, W., M. Ross, R. Shackleton, and P. Wilcoxen. 1999. *Emissions Trading, Capital Flows and the Kyoto Protocol*. Discussion Paper in International Economics 144. Brookings.

National Board of Trade (Kommerskollegium), Government of Sweden. 2004. *Climate and Trade Rules: Harmony or Conflict?*

Pauwelyn, Joost. 2007. "U.S. Federal Climate Policy and Competitiveness Concerns: The Limits and Options of International Trade Law." Nicholas Institute for Environmental Policy Solutions Working Paper 0702. Duke University.

Reinaud, Julia. 2008. "Issues Behind Competitiveness and Carbon Leakage – Focus on Heavy Industry." IEA Information Paper. International Energy Agency: Paris.

Sampson, Gary. 1999. "WTO Rules and Climate Change: The Need for Policy Coherence." In *Global Climate Governance: A Report on the Inter-Linkages between the Kyoto Protocol and Other Multilateral Regimes*, ed. Bradnee Chambers. United Nations University.

———. 2000. *Trade, Environment and the WTO: The Post-Seattle Agenda*. Policy Essay 27. Washington: Overseas Development Council..

Seidman, Laurence, and Kenneth Lewis. 2009. "Compensation and Contributions under an International Carbon Treaty." *Journal of Policy Modeling* 31, no. 3: 341-350.

Slater, J. 2008. "Climate Change: Competitiveness Concerns and Prospects for Engaging Developing Countries." Testimony before the Energy and Air Subcommittee, Energy, and Commerce Committee, U.S. House of Representatives, March 5.

U.S. Energy Information Administration. 2006. *Energy Market Impacts of Alternative Greenhouse Gas Intensity Reduction Goals*. Report SR/OIAF/2006-01. U.S. Department of Energy.

———. 2007. *Emissions of Greenhouse Gases in the United States 2006*. Report DOE/EIA-0573(2006). U.S. Department of Energy.

———. 2008. *Energy Market and Economic Impacts of S. 2191, the Lieberman-Warner Climate Security Act of 2007*. Report DOE/EIA-SR/OIAF/2008-01. U.S. Department of Energy.

Wall Street Journal. 2008. "Unions Back Carbon Tax on Big Polluting Nations." January 16.

Webster, D. G. 2008. *Adaptive Governance: Dynamics of Atlantic Fisheries Management*. MIT Press.

Weinstein, M. 2001. "Greens and Globalization: Declaring Defeat in the Face of Victory." *New York Times*, April 22.

THOMAS L. BREWER 4

Technology Transfers and Climate Change: International Flows, Barriers, and Frameworks

Discourse concerning international technology transfers to address climate change issues is typically based on a paradigm that is focused on North-South technology flows and financial flows, especially in the context of official development assistance programs or offset projects under the Clean Development Mechanism of the Kyoto Protocol. This paradigm is useful for many analytic and negotiating agendas. However, it reflects an overly narrow conceptualization of the nature, sources, and methods of international technology transfers. It thus neglects important issues that need to be addressed in order to utilize more fully the potential of international technology transfers for climate change mitigation and/or adaptation.

This chapter proposes a complementary paradigm, which emphasizes the importance of trade and foreign direct investment as the principal mechanisms by which technology is transferred internationally. The paradigm reflects diversity in the geography of technology flows, a different focus on the types of international economic flows that facilitate technology transfers, a different set of barriers to international technology flows, and different institutional frameworks that can facilitate or impede technology flows.[1]

The second section of this chapter presents these paradigms in a more detailed and comparative discussion.[2] The third section considers evidence about the geography of international technology flows, with an explicit recog-

The author is indebted to the following for their comments on the paper at the conference: Brian Flannery, Kelly Sims Gallagher, Muthukumara Mani, and Jacob Werksman. Gerry T. West also made useful comments on the conference paper before it was revised. Several changes were made to reflect these comments in revising the paper for publication here.

1. Barton (2007) has also noted a shift in paradigms of technology transfer more generally—from a paradigm focused on licensing and intellectual property rights to a paradigm focused on foreign direct investment.

2. A subsequent conference paper (Brewer 2008a) develops a third paradigm, based on international public-private cooperation agreements. That research was prompted in part by the insightful comments of Jacob Werksman at the Brookings Institution conference. Further research on international institutional issues is being conducted on international public-private partnerships.

nition of developing countries as sources of such flows. The fourth section analyzes evidence about the barriers to international technology flows. The fifth section describes the international institutional frameworks affecting the flows. The sixth section discusses the implications of these paradigms and flows for international negotiations.

Paradigms of International Technology Transfer

The term "technology" continues to be used in common parlance to refer mostly to tangible goods, for instance, computer hardware; but it is typically used in the academic literature to include more intangible elements of organizations' activities as applied knowledge or know-how. Further, the term is used to include managerial know-how, not only in the engineering of production processes but also more generally in management processes.

Over time, the literature on international technology transfers has progressed from a relatively narrow definition of technology as "scientific and engineering knowledge, . . . which . . . are principally the outcome of R&D [research and development]. The transfer of this codified knowledge constituted technology transfer" in the earlier narrow notion of technology.[3] Current notions of technology, however, include a second notion as well—technology as *tacit* knowledge that is embedded in firms' procedures and personnel. Though the first conceptualization leads to an analytic focus on explicit knowledge concerning specific products and their associated production processes, the second conceptualization leads to a focus on the capabilities and processes of firms, especially the tacit knowledge that is embedded in them. This encompassing notion of technology that includes both its "soft" and "hard" manifestations is now widespread in government policymaking and business strategy circles as well as academic circles, and it is reflected in the Stern Report on the economics of climate change.[4]

Relevant Technologies for Climate Change Mitigation or Adaptation

There are numerous lists of industries and products (both goods and services) related to climate change.[5] A well-known list was developed by Pacala and

3. Cantwell (2009, 420).
4. Stern (2007, 567); also see Brewer (2008b) and Fischer, Egenhofer, and Alessi (2008).
5. When identifying technologies in terms of standardized product or industry classification schemes—such as the Harmonized System of the World Customs Union or the International Standard Industrial Classification system or the UN Product Code—there are a variety of technical issues, which can be important in negotiations. These and related issues have been raised recently, especially with regard to international technology transfer and thus trade issues; see Howse and

Table 4-1. Types of Greenhouse Gas Emission–Reducing Technologies

---

Reducing emissions from energy supply and infrastructure

Low emission, fossil-based power and fuels

    Zero-emission power, hydrogen, and other value-added products

    High-efficiency coal / solid feedstock

    High-efficiency gas fuel cell / hybrid power systems

Hydrogen

    Hydrogen production from nuclear fission and fusion

    Integrated hydrogen energy systems

    Hydrogen production

    Hydrogen storage and distribution

    Hydrogen use

    Hydrogen infrastructure safety

Renewable energy fuels

    Wind energy

    Solar photovoltaic power

    Solar buildings

    Concentrating solar power

    Biochemical conversion of biomass

    Thermochemical conversion of biomass

    Biomass residues

    Energy crops

    Photoconversion

    Advanced hydropower

    Geothermal energy

Nuclear

    Existing plant research and development

    Next-generation fission energy systems

    Near-term nuclear power plant systems

    Advanced nuclear fuel cycle processes

    Nuclear fusion

Energy infrastructure

    High-temperature superconductivity

    Transmission and distribution technologies

    Distributed generation and combined heat and power

    Energy storage

    Sensors, controls and communications

    Power electronics

Reducing emissions from energy use

Transportation

    Light vehicles—hybrids, electric, and fuel cell vehicles

    Alternative-fueled vehicles

    Intelligent transportation systems infrastructure

    Aviation

    Transit buses-urban duty-cycle

Table 4-1. Types of Greenhouse Gas Emission–Reducing Technologies (continued)

Buildings
> Building equipment, appliances, and lighting
> Building envelope (insulation, walls, roof)
> Intelligent building systems
> Urban heat island technologies

Industry
> Energy conversion and utilization
> Resource recovery and utilization
> Industrial process efficiency
> Enabling technologies for industrial processes

Enhancing capabilities to measure and monitor emissions
Hierarchical measuring and monitoring systems
> For energy efficiency
> For geologic carbon sequestration
> For terrestrial carbon sequestration
> For ocean carbon sequestration
> For other greenhouse gas

Reducing the climate effect of non–carbon dioxide greenhouse gas emissions
Methane emissions from energy and waste
> Anaerobic and aerobic bioreactor landfills
> Conversion of landfill gas to alternative uses
> Electricity generation technologies for landfill gas
> Advances in coal mine ventilation air systems
> Advances in coal mine methane recovery systems
> Measurement and monitoring technology for natural gas systems

Methane and nitrous oxide emissions from agriculture
> Advanced agriculture systems for nitrous oxide emissions reduction
> Methane reduction options for manure management
> Advanced agriculture systems for enteric emissions reduction

Emissions of high global-warming potential gases
> Semiconductor industry; abatement technologies
> Semiconductor industry: substitute for processes producing gases with high global-
>   warming potential
> Semiconductor and magnesium: recovery and recycle
> Aluminum industry: perfluorocarbon emissions
> Electric power systems and magnesium: substitute for SF_e
> Supermarket refrigeration: hydrofluorocarbon emissions

Nitrous oxide emissions from combustion and industrial sources
> Nitrous oxide abatement technologies from nitric acid production
> Nitrous oxide abatement technologies for transportation

Emissions of tropospheric ozone precursors and black carbon
> Abatement technologies for emissions for tropospheric ozone precursors and
>   black carbon

Source: UNFCCC (2002, 63).

Socolow on the basis of the fifteen "wedges" that they identified as having the potential to contribute a reduction of 1 gigaton of carbon dioxide equivalent per year by 2054.[6] The United Nations Framework Convention on Climate Change (UNFCCC) has compiled a lengthy list, as presented in table 4-1.[7]

These lists of mitigation technologies differ in their level of detail and hence technological specificity, and they also differ in whether they include or exclude nuclear power in particular. At an aggregate level, however, these lists share many of the same technologies, especially those concerning energy efficiency and renewable energy sources, which are the focus of this chapter. Of course, the technologies vary in their cost-effectiveness in reducing greenhouse gas (GHG) emissions and in the scale of deployments required for them to have notable effects.[8] Among them, carbon capture and sequestration is often mentioned as the most important, particularly with regard to coal-fired power plants; however, because its commercialization is still perhaps a decade or even more in the future, it is not included in this chapter's analysis of flows and barriers to them. Otherwise, however, energy efficiency technologies, renewables such as wind and solar, and biofuels, which are all widely mentioned as key technologies, are included in the present analysis. These key technologies are evident in the many lists that have been developed by numerous organizations and authors. As noted further below, however, there are significant variations in the appropriateness of some technologies, depending on the level of economic and technological development of the recipient countries.

Stern provides a list of nine types of technologies that could reduce carbon emissions in the energy sector: efficiency, carbon capture and storage, nuclear, biofuel, combined heat and power, solar, wind, and hydropower.[9] An Environmental Technologies Action Plan developed by an advisory group of the European Commission includes fifty-one categories organized in a matrix based on two dimensions.[10] One dimension is industry sector—for instance, energy supply—whereas the other consists of a combination of energy efficiency and renewables, carbon sequestration, and hydrogen and fuel cells. The U.S. Climate Change Technology Program itemizes hundreds of technologies, which

---

van Bork (2006); IEA/OECD (2006); Sugathan (2006); and World Bank (2007). Furthermore, at the World Trade Organization, classification issues are different for goods and services because the agreement on goods (GATT) and the agreement on services (GATS) use different product classification schemes (Brewer, forthcoming).

6. Pacala and Socolow (2004); also see Socolow (2006) and Socolow and Pacala (2006).

7. UNFCCC (2006b).

8. The sources for GHG emissions (Vattenfall 2007) and for effects (IEA 2008) provide extensive analyses in easy-to-compare formats of the wide range of technologies that are available now or in the next few decades.

9. Stern (2007, 259),

10. European Commission (2003).

are listed in "current portfolios" and "future research directions."[11] They are organized according to end-use/infrastructure (for example, transportation), energy supply (hydrogen), carbon capture-storage (geologic storage), non–carbon dioxide GHGs (methane from landfills), and measuring and monitoring capabilities (oceanic carbon dioxide sequestration). A list of renewable energy goods from the U.S. Trade Representative is more narrowly focused.[12] A list of technologies developed by Lehman Brothers investment bank includes renewables (hydropower, geothermal power, wind, solar, ocean, and biomass), carbon capture and storage, integrated gasification combined cycle, and nuclear power.[13] The World Bank has developed a list of technologies for energy production, supply, and end use that includes many of the same technologies as the other lists.[14]

The Technology Needs Assessments conducted in developing countries for the UNFCCC Secretariat provide both much commonality and much variation in the kinds of technologies that are viewed as important. A summary of the reports for over twenty countries found that the following were among the "key technologies": "renewable energy for small-scale applications, such as biomass stoves; combined heat and power; and energy-efficient appliances and building technologies such as compact fluorescent light bulbs. For transport, traffic management and cleaner vehicles for public transport were most important."[15] Additional technologies and variations among countries are evident in individual country reports. The top priorities for Ghana, for instance, included industrial energy improvements (including boiler efficiency enhancement), methane gas capture from landfill sites, biofuels (from jatropha, an oilseed plant), and energy-efficient lighting (compact fluorescent bulbs).[16]

Some countries have developed extensive, industry-specific needs, which of course also vary across countries. Vietnam, for instance, lists ten quite specific GHG-reducing technologies that are needed in its cement industry—among them, "large vertical roller mill for raw material crushing, . . . vertical roller mill for coal crushing, . . . high-efficiency separator in the finishing process, . . . burning used tires as substitute for cement kiln." As of April 2008, there were thirty-nine country reports in various stages of completion, some of them extremely detailed; Indonesia's, for instance, was 299 pages long. Individually and collectively, they offer developing country perspectives that are in some respects quite different from those of the developed countries. Yet, at the same

11. U.S. Climate Change Technology Program (2006).
12. U.S. Office of the Trade Representative (2006).
13. Lehman Brothers (2007).
14. World Bank (2006a, 10).
15. Stern (2007, 564–65).
16. Stern (2007, 565).

time, there are many similarities. See for instance, a summary table of "the promise of technology" in a UNFCCC report.[17]

Although such lists tend to identify "new" or otherwise "alternative" technologies, it should not go unnoticed that "old" technologies are also relevant. For instance, turboprop aircraft engines are much more fuel efficient than jet engines. As a result, orders for medium-range turboprop planes have been increasing substantially, particularly by airlines in Europe, which have used such technology extensively for many years but had been gradually replacing the turboprop planes with jet planes. In the maritime shipping industry, there is interest in a step "backward" to readapt the use of wind power on large merchant vessels by the use of large sails; a vessel thus equipped as a pilot project has made a long-distance ocean voyage.

Climate change *adaptation* technologies have not yet been so systematically or explicitly identified, nor have they been the subject of such extensive interest, compared with mitigation technologies. However, the UNFCCC has identified a diverse variety of technologies for adaptation, including for coastal zones, water supplies, agriculture, and health.[18]

Beyond common notions of technology and an openness about the relevant technologies for mitigation and/or adaptation, the two paradigms of climate change technology transfer are quite different in their principal features. It is useful to briefly examine each one.

Paradigm I: North-South Technology and Financial Flows

Paradigm I, which is currently dominant, is reflected in numerous documents produced over time in the implementation and negotiating processes of the UNFCCC.[19] Although there has been some increased interest in South-South transfers,[20] the predominant focus remains on North-South technology transfers and North-South financial transfers, particularly through bilateral and multilateral official development assistance (ODA) channels.

Paradigm II: Global Technology, Trade, and Investment Flows

Paradigm II exhibits key differences from Paradigm I. First, it reflects the fact that there are significant technology transfers among many groups of countries, including from developing countries to developed countries as well as to other developing countries, and of course among developed countries. Second,

17. UNFCCC (2004, 63).
18. UNFCCC (2006c).
19. UNFCCC (2002, 2006a, 2006c, 2007a).
20. For example, see South Africa (2006).

Table 4-2. Summary Comparison of Paradigms

Elements of Paradigm	Paradigm I: North-South Technology and Financial Flows	Paradigm II: Global Technology, Trade, and Investment Flows
Flows: geography	North to South	South-North South-South North-North North-South
Flows: units of analysis	Countries' financial flows	Firms' trade and investment in goods and services
Barriers	Technological-administrative capacities and intellectual property rights policies in developing countries	Trade and investment policies in all countries
Institutional frameworks	International development banks and bilateral official development assistance	Multilateral, regional, and bilateral trade and investment agreements and organizations

Source: Developed by the author.

the technology flows occur, in substantial part, as a result of trade and foreign direct investment (FDI) flows.

Comparing the Paradigms

Table 4-2 provides a side-by-side summary comparison of the two paradigms. Of course, the conceptualizations are useful only insofar as they reflect empirically valid representations of reality and/or offer relevant policy prescriptions. Thus, the remaining sections of the chapter turn to evidence concerning the flows, barriers to them, institutional frameworks, and then the policy implications.

Patterns of International Flows

The data given in table 4-3 provide aggregate perspectives on key features of international investment flows. First, the relative magnitudes of ODA flows remain quite small. (This fact, of course, is a basis of the Paradigm I policy prescription for more ODA.) Bilateral ODA and multilateral ODA in 2000 were respectively only 0.7 and 0.4 percent of total gross fixed capital formation in non–Annex I countries.[21] Second, international direct investment and interna-

21. Annex I is an annex in the UNFCCC. "Annex I" countries are those that committed themselves as a group to reducing their emissions of the six greenhouses gases by at least 5 percent

Table 4-3. Sources of Global Investment in Gross Fixed Capital Formation (GFCF)

Sector	FDI Flows (percent)	International Borrowing (percent)	Bilateral ODA (percent)	Multilateral ODA (percent)	Domestic (percent)	Total GFCF (billions of dollars)
All sectors	17.9	17.2	0.1	0.1	64.7	7,750
Agriculture, hunting, forestry, fisheries	1.0	5.4	0.3	0.2	93.1	175
Mining	33.2	0.2	0.2	0.4	66.4	139
Manufacturing	22.1	6.0	0.0	0.0	71.9	1,301
Electricity, gas, water	12.2	16.4	1.7	0.9	68.8	257
Transport, storage, communications	16.7	16.8	0.5	0.4	65.5	889
In Annex I countries only						
All sectors	12.5	19.4	0.0	0.0	68.1	6,014
Agriculture, hunting, forestry, fisheries	0.0	8.9	0.0	0.0	91.1	104
Mining	0.8	0.3	0.0	0.0	98.9	68
Manufacturing	N.A.	N.A.	N.A.	N.A.	N.A.	858
Electricity, gas, water	0.0	18.5	0.0	0.0	81.4	186
Transport, storage, communications	0.3	22.2	0.0	0.0	77.5	630
In non–Annex I countries only						
All sectors	10.2	3.8	0.7	0.4	85.0	1,654
Agriculture, hunting, forestry, fisheries	1.7	0.2	0.8	0.5	96.9	68
Mining	17.8	0.0	0.4	0.1	98.9	69
Manufacturing	15.3	0.5	0.1	0.0	84.1	443
Electricity, gas, water	12.6	5.8	0.6	3.3	77.7	67
Transport, storage, communications	8.9	1.5	1.7	1.4	86.4	248

Source: Excerpted by the author from UNFCCC (2007, tables 35.1–31.7). Data are for 2000.
Note: FDI = foreign direct investment. ODA = official development assistance. Annex I is an annex in the UNFCCC. The "Annex I" countries are those that committed themselves as a group to reducing their emissions of the six greenhouses gases by at least 5 percent below 1990 levels from 2008 to 2012. Specific targets vary from country to country. The "non–Annex I countries" are developing countries that are not listed in Annex I and thus have no emissions reduction targets (from www.unfccc.org). N.A. = not applicable.

tional debt are both on the order of hundreds of billions of dollars per year, compared with total ODA flows on the order of hundreds of millions.[22] Third, international investments are overwhelmingly flowing into Annex I countries, with non–Annex I countries receiving only about 9 percent of the total in 2000.

below 1990 levels from 2008 to 2012. "Non–Annex I countries" are developing countries that are not listed in Annex I and thus have no emissions reduction targets.

22. There are numerous conceptual and empirical issues about the precision of data about FDI and other investment flows—issues that are beyond the scope of this chapter. For present purposes, these data are useful as indicators of approximate relative magnitudes. Readers who are interested in detailed discussions of these data-quality issues should consult the annual reviews of international direct investment by UNCTAD and world debt by the World Bank.

Finally, international flows are about one-third of total world gross fixed capital formation, compared with two-thirds from domestic sources.[23]

The sectoral distributions are also evident in table 4-3. For non–Annex I countries, in particular, the relative importance of FDI flows—in mining; manufacturing; electricity, gas, and water; and transport, storage, and communications—is evident. Whereas FDI flows are between 9 and 18 percent in each of those sectors, ODA is only 4 percent at most, and less than 2 percent in four of the five sectoral categories.

As for sustainable energy sectors, data from the United Nations Environment Program indicate that in 2006, three industries received most of the investment—wind (27 percent), biofuels (18 percent), and solar (11 percent)—with other renewables and energy efficiency receiving less than 5 percent each.[24] Relatively large proportions of world investment in sustainable energy are made in Europe and the United States, which together accounted for about $50 billion, or two-thirds of the world total, in 2006. China accounted for about $6 billion, and India about $4 billion. Of course, projections of the geographic and sectoral patterns and trends are of special interest for the future of climate change mitigation and adaptation.[25]

Developing Countries as Technology Sources, and South-South Transfers

Table 4-4 reveals that several developing countries are world leaders in a variety of key climate friendly technologies. They include not only China in numerous technologies (coal gasification, compact fluorescent lightbulbs, solar voltaic cells, and wind power) but also South Africa in coal–to–synthetic fuels technologies, and Mexico in solar hot water heaters. As for biofuels, although recent studies have undermined claims about the net GHG-reducing effects of some ethanol sources, especially when made with U.S.-produced corn, there is continuing interest in Brazil's world leadership in the production of ethanol using sugarcane, particularly if and when it does not involve the destruction of the rain forest to create the sugar plantations.

India is well known already for its wind power industry, as noted in more detail below. Its leadership in the use of jatropha tree berries as a second-generation feedstock for biodiesel production is less well known and still in its infancy; India has already been involved in numerous projects for transferring

23. Because some foreign direct investments are partially financed from local host country sources, the distinction between international and domestic sources, as revealed in balance of payments statistics, is not entirely accurate.

24. UNEP (2007, figure 5).

25. These projections can be found in reports of the IEA (2008), UNFCCC (2007b), World Bank (2006b, 2007), and Cambridge Energy Research Associates (2008).

Table 4-4. Developing Country Leaders in Climate Friendly Industries and Products

Industries and Products	Developing Country Leaders
Biofuel (ethanol)	Brazil (sugarcane feedstock and refining processes)
Biofuel (biodiesel)	India (jatropha, next-generation feedstock)
Coal gasification	China
Coal to synthetic fuels	South Africa
Coal to hydrogen	China
"Clean" coal	Mexico
Compact fluorescent lamps	China, Indonesia
Heat pumps	China
Solar photovoltaic cells	China (third leading producer, after Germany and Japan)
Solar hot water heaters	Mexico
Wind energy	China, India

Source: Compiled by the author from Socolow (2006) and World Bank (2008).

jatropha-based biodiesel technology to other developing countries, including Ghana, Indonesia, Mozambique, and the Philippines.[26]

Other examples of South-South transfers of climate change–mitigating technologies abound. For example, China has assisted in the construction of over seventy small-scale biogas digesters for manure decomposition–methane production in twenty-two countries and provided training in their use. And ceramic cookstoves that provide 20 to 50 percent efficiency gains over conventional charcoal stoves were developed based on a collaboration between firms in Kenya and Thailand and have been used by more than 700,000 homes in Kenya, as well as in other African countries.[27]

Foreign Direct Investment and Multinational Firms

Paradigm II not only includes technology flows emanating from non–Annex I developing countries—whether to other developing countries or to Annex I countries—it also posits that technology flows are typically embedded in the trade and FDI flows of firms, particularly large multinational firms. Some of these multinational firms—indeed, an increasing number—are themselves based in developing countries.

Two such firms based in India are notable examples. One is Suzlon, which is a world leader in wind turbine and gearing technologies and has foreign subsidiaries and other affiliates in the United States and several European countries, as well as China.[28] The other Indian firm is Tata, the well-known diversified

26. For details see *FO Licht's World Ethanol & Biofuels Report* (Agra Europe 2006 and 2007).
27. For details on these and other examples see UNFCCC (2006b).
28. For details see Alavi (2007, 15) and Lewis (2007, table 2).

conglomerate with extensive business interests in energy, transportation, and other climate relevant industry sectors. It has wholly owned subsidiaries, joint ventures, and other forms of international business relationships for R&D and manufacturing in more than two dozen countries.[29]

Of course, there are many other large multinational firms with significant roles in the development and international dissemination of leading-edge technologies in the wind power industries and more generally in the energy sector, and they are mostly headquartered in Annex I countries. General Electric and Siemens are two well-known examples. What is especially noteworthy about them in the present context is that they both have extensive climate change–related energy R&D activities in China and India.

More generally, all these firms—Suzlon, Tata, General Electric, and Siemens—illustrate a basic fact about international technology transfer: It takes place to a great extent within multinational firms' internationalized R&D and manufacturing processes, sometimes with collaborators from local host countries. In fact, the research-development-diffusion process for any one given technology often involves many firms in many countries.

Although these examples are large multinationals, it should not be assumed that large firms are the only important sources of technology innovation or international diffusion. Indeed, small and medium-sized firms are often the originators of new technologies and even the principal internationalizers of them. For instance, small and medium-sized firms have been instrumental in the development of biodiesel technology in many countries. A current example is the development of hybrid biodiesel-electric pickup trucks by firms in India and the United States.

Further, it should be noted that large multinational firms are often accused of limiting the effective transfer of technologies into local foreign economies; they often prefer to keep key technologies internalized with the firm—including within their foreign affiliates—rather than allow their technologies to be externalized, because such technologies are frequently a source of these firms' competitive advantage over their rivals. In fact, the *internalization* of these technologies is a core concept in the prevailing theories of multinational firms and FDI.[30] Multinational firms are, therefore, themselves sometimes barriers to the international transfer of technologies.[31]

---

29. For details see Tata website (www.tata.com [November 2006]).

30. Brewer and Young (2000, 2009); Cantwell (2009).

31. Although intellectual property rights are sometimes used as a formal way to protect explicit proprietary technology, the internalization of explicit and tacit technologies within the managerial processes of firms is a more significant barrier.

Government Policy Barriers to International Flows

The nature, availability, and quality of the evidence about government policy barriers to international technology transfers is highly variable. Tariff data on trade in goods are the most readily available and the most easily comparable across countries and over time because of the widely used Harmonized System developed by the World Customs Organization and used in World Trade Organization (WTO) negotiations. Nontariff barriers (NTBs) to trade in goods are more problematic to identify and measure, but they can be estimated in terms of their tariff equivalents, as illustrated further below. Measuring the impact of barriers to international *service* transactions is yet more challenging because there is less consistency in the categories used to describe such transactions, though the WTO and the United Nations have both used service industry classification systems. Finally, barriers to FDI are also sometimes difficult to identify and measure, though they are subject to annual reviews by the United Nations Conference on Trade and Development in its *World Investment Report*.[32]

Of these many forms of barriers to international technology transfer for climate change mitigation or adaptation, this chapter considers tariffs—and, to a lesser extent, NTBs—as they pertain to trade in goods. Table 4-5 contains evidence about the tariff levels affecting the energy efficiency of buildings. It is clear that the tariff levels of the three non–Annex I countries (Brazil, India, and Mexico) are generally higher than those of the Annex I countries (Canada, European Union members, Japan, and the United States). Except for Canadian tariffs of 15.1 percent on walled insulating units of glass, in every instance the non–Annex I countries' tariffs are higher. Of course, these are not only obstacles to North-South technology transfers, they are also obstacles to South-South transfers.

Impediments to trade in climate friendly products among developing countries are more explicitly shown in table 4-6. In this table, the technologies and their developing country sources are identified in the top two rows, and the levels of the tariffs and NTBs to importing them in five developing countries are indicated in the lower rows. With few exceptions, the tariffs are in the two-digit range, and the NTBs, expressed as tariff equivalents, are in the two- to three-digit range. Thus, there are substantial barriers to South-South technology transfer through trade in these climate friendly goods.

EU tariffs of as much as 66 percent on compact fluorescent lightbulbs (CFL) imported from China are an especially noteworthy and timely example of bar-

32. The 2007 report, the most recent available at the time of writing, is available at the website of the United Nations Conference on Trade and Development (www.unctad.org/Templates/Download.asp?docid=9001&lang=1&intItemID=4361 [March 2009]).

Table 4-5. Illustrative Tariffs Affecting Energy Efficiency of Buildings

	Tariff Rate (percent)			
Country or Group	Multiple-Walled Insulating Units of Glass (HS 7019.90)	Other Glass-Fiber Products (HS 8419.50)	Fluorescent Lamps, Hot Cathode (HS 8539.31)	Thermostats (HS 9032.10)
Non–Annex I countries				
Brazil	12.0	18.0	18.8	18.0
India	60.0	35.0	100.0	60.0
Mexico	11.7	9.2	15.0	15.0
Annex I countries				
Canada	15.1	1.2	9.7	5.7
European Union	8.0	1.2	3.6	2.8
Japan	0.0	0.0	0.0	0.0
United States	5.8	3.4	3.2	3.6

Source: Compiled by the author from OECD (2001, annex 3); and Steenblik (2006, 54–55, 70).
Note: Tariff rates are applied most-favored-nation rates, in effect c. 2000; HS = Harmonized System of Tariff Classifications.

riers to South-North trade in a climate friendly technology. Although rationalized on antidumping grounds, these tariffs led to a precipitous decline in Chinese CFL exports to Europe and the concomitant collapse of numerous Chinese CFL manufacturers.[33]

U.S. limits on tax credits for purchasers of hybrid fuel automobiles is an example of a barrier to North-North transfer of a climate friendly technology. The United States imposed de facto, firm-specific quotas (without explicitly naming the firms) on the number of hybrid fuel automobiles produced by any one manufacturer that could receive a tax credit—quotas that happen to affect only Japan-based manufacturers, because their hybrids are the best-selling models in the United States.

Table 4-7 indicates for the wind energy industry that barriers to trade in services and to FDI, along with tariffs on trade in goods, can be significant obstacles to international technology transfers. Further, the types of products (whether goods or services) interact with the types of barriers (whether specially targeted, for defined products, or generic across products).

In the WTO, governments' commitments on trade and investment liberalization, as specified in the General Agreement on Tariffs and Trade (GATT) and the General Agreement on Trade in Services (GATS), respectively, are indicated in terms of products. This approach has been challenged by India in its response to the EU–United States proposal for eliminating tariffs on climate-friendly goods; instead of *product-specific* liberalization, India has proposed

33. The tariffs were in effect for seven years, from mid-2001 until late 2008.

Table 4-6. Tariffs and Nontariff Barriers (NTBs) on Goods in Four Select Product Categories

	Product Category							
	Compact Fluorescent Lamps *China, Indonesia*		Wind Energy *China*		Solar Voltaic *China, Taiwan, Malaysia, South Korea*		Lower-Carbon Coal *Mexico*	
	Developing Countries among World's Top Ten Exporters							
Importing Country	*Average Applied Tariff (percent)*	*Average Applied NTBs Tariff Equivalent (percent)[a]*	*Average Applied Tariff (percent)*	*Average Applied NTBs Tariff Equivalent (percent)[a]*	*Average Applied Tariff (percent)*	*Average Applied NTBs Tariff Equivalent (percent)[a]*	*Average Applied Tariff (percent)*	*Average Applied NTBs Tariff Equivalent (percent)[a]*
Brazil	18	96	14	14	18	53	14	145
China	8	N.A.	8	15	10	N.A.	15	25
India	15	102	15	15	15	41	15	N.A.
Mexico	15	N.A.	15	12	13	62	12	N.A.
South Africa	N.A.	N.A.	N.A.	0	12	N.A.	0	125

Source: Compiled by the author from World Bank (2008).

Note: The Harmonized System (HS) codes for the product groups are as follows: lower carbon coal technologies, HS codes 840510, 840619, 841181, 841182, 841199; energy-efficient lighting, HS code 853931; wind energy, HS codes 848340, 848360, 850230; and solar photovoltaic systems, HS codes 850720, 853710, 853931, 854l409. N.A. = not applicable.

a. The data on NTBs are derived from the World Bank's trade database. The NTBs were calculated by transforming all the information on NTBs into a price equivalent. The ad valorem equivalent of the core NTBs thus calculated includes price and quantity control measures, technical regulations, and monopolistic measures, such as a single channel for imports.

Table 4-7. Interactions of Wind Energy Products and Barriers to International Trade, Investment, and Technology Transfer

Type of Product: Goods and Services	Type of Barrier to International Trade, Investment, and Technology Transfer		
	Specifically Targeted	Defined Categories for Product Groups	Generic across Product Groups
Specific good or service: windmill generators	Tariffs on imported windmill generators	Tariffs on blades of various types	FDI local joint venture requirements for manufacturing facilities
Dual use: electrical wires	Tariffs on electrical wires	Safety certification requirements for electrical goods	Import inspection policies
General: construction engineering services	Restrictions on cross-border trade in construction engineering services	Licensing of engineering services firms	Restrictions on movement of natural persons

Source: Developed by the author from information in Barton (2007, 20–30); also see Lewis (2007).

project-specific reductions because many of the products are dual use and thus can be used for purposes other than climate change mitigation.[34] However, there has been an impasse over this suggestion, because the EU and the United States continued to favor a product-specific approach. A compromise consisting of a combination was also being considered.

Although the emphasis in the present analysis is on government international trade and investment policy barriers to international technology transfers, it should also be noted that "domestic" government regulations, subsidies, and other policies are also important determinants of technology flows. The significance of such policies—or their absence, in some instances—as incentives or disincentives is emphasized in Gallagher's study of the automotive industry in China.[35]

International Institutional Frameworks

There is already a wide range of both climate change and trade-investment international institutional arrangements that affect technology transfers, and they exist at the bilateral, regional, and plurilateral levels as well as the mul-

34. See the International Centre for Trade and Sustainable Development publication (ICTSD 2006) and especially the contributions by Sell (2006) and Sugathan (2006).
35. Gallagher (2006).

Table 4-8. Existing International Institutional Arrangements Concerning Climate Change and Trade

Type of Arrangement	Climate Change	Trade and Investment
Multilateral	UN Framework Convention on Climate Change Kyoto Protocol Other UN agencies International financial institutions Global Environment Facility	World Trade Organization: General Agreement on Tariffs and Trade General Agreement on Trade in Services Trade-Related Intellectual Property Rights Dispute Settlement Understanding
Regional	Asia-Pacific Partnership	Asia-Pacific Economic Cooperation, European Union, Mercosur, North American Free Trade Agreement, others
Bilateral	United States–Brazil, Sweden, others European Union, others Japan, others	Free trade agreements, bilateral investment agreements
Plurilateral/sectoral Aviation Shipping Others	Pending Pending	International Civil Aviation Organization International Maritime Organization

Source: Compiled by the author from various documents of the UNFCCC, WTO, other organizations and governments.
Note: These are not exhaustive lists.

tilateral level (see table 4-8). The analysis here focuses on the multilateral institutions.

To date, the multilateral institutional arrangements for the climate regime manifest the emphases of Paradigm I on technology flows and financial flows from North to South. This is evident in particular in the work of the UNFCCC Expert Group on Technology Transfer, whose mandate is to facilitate the implementation of Article 4, paragraph 5, of the UNFCCC:

> The developed country Parties and other developed Parties included in Annex II shall take all practicable steps to promote, facilitate and finance, as appropriate, the transfer of, or access to, environmentally sound technologies and know-how to other Parties, particularly developing country Parties, to enable them to implement the provisions of the Convention. In this process, the developed country Parties shall support the development and enhancement of endogenous capacities and technologies of developing

country Parties. Other Parties and organizations in a position to do so may also assist in facilitating the transfer of such technologies.[36]

Similar emphases are evident in the Bali Action Plan, which calls for

... (d) Enhanced action on technology development and transfer to support action on mitigation and adaptation, including, inter alia, consideration of: (i) Effective mechanisms and enhanced means for the removal of obstacles to, and provision of financial and other incentives for, scaling up of the development and transfer of technology to developing country Parties in order to promote access to affordable environmentally sound technologies; (ii) Ways to accelerate deployment, diffusion and transfer of affordable environmentally sound technologies; (iii) Cooperation on research and development of current, new and innovative technology, including win-win solutions; (iv) The effectiveness of mechanisms and tools for technology cooperation in specific sectors.[37]

Furthermore, there is increasing interest in the nature and extent of technology transfers associated with Clean Development Mechanism and Joint Implementation projects,[38] and the effects of host country investment and regulatory policies on the features of such projects.[39]

The Global Environment Facility, the various carbon programs at the World Bank, and the new Clean Energy Fund to be administered by the World Bank are also all important elements of the current multilateral system for funding technology transfers to developing countries. Because they are all likely to become much larger in the next few years, they will become increasingly significant, and they can be used to leverage other funding sources. Yet the amounts of funds are likely to remain small in relationship to the technology flows associated with trade and investment through private channels.

Paradigm II, of course, implies that the international institutional arrangements *for trade and investment* are also relevant to international technology transfers for climate change mitigation and adaptation. An extensive analysis of the complexities of those arrangements would be far beyond the scope of this chapter.[40] However, it can be briefly noted that the coverage is highly uneven across products (goods or services), methods of technology transfer (trade or FDI), and the geographic scopes of the agreements (multilateral, plurilateral, regional, bilateral). At the WTO, in particular, the coverage of methods of technology transfer varies significantly between the GATT for goods and

36. The work of the Expert Group on Technology Transfer is discussed in detail in UNFCCC (2006a); its first work program is summarized in UNFCCC (2002), and its most recent work program is summarized in (2007a). Its clearinghouse for technology transfer information is accessible at www.ttclear.unfccc.int.

37. This is from www.unfccc.int (March 2008).

38. DeConinck, Haake, and van der Linden (2007).

39. Anger, Bohringer, and Moslener (2007).

40. For the details, see Brewer (2008c).

**Table 4-9. Methods of International Technology Transfer, Climate Friendly Technology
Examples, and Coverage in the World Trade Organization**

Methods	Goods	Services
Production in exporting country / consumption in importing country	GATT: tariffs and nontariff barriers	GATS: "consumption abroad"
Foreign direct investment, including joint ventures	GATT/TRIMs only	GATS: "commercial presence" (that is, foreign direct investment)
Temporary relocation of employees	Not covered	GATS: "movement of natural persons"
International migration of skilled people	Not covered	Not covered
Licensing	TRIPS	TRIPS

Note: GATT = General Agreement on Tariffs and Trade; GATS = General Agreement on Trade in Services; TRIMs = Trade-Related Investment Measures; WTO = World Trade Organization; TRIPS = Trade-Related Intellectual Property Rights.

the GATS for services (see table 4-9). Further, because of the limited coverage of FDI in the WTO, the FDI provisions of regional and bilateral agreements are particularly important, and there are many hundreds of them.[41]

Implications for International Institutional Architectures

The negotiating agendas in both the international climate change and trade-investment arenas also need to take into account the broad array of barriers to international technology transfers for climate change mitigation or adaptation.[42] Multilateral and bilateral official development assistance levels need to be increased significantly. But that is not enough, because most technology transfers occur through private international trade and FDI transactions, and they encounter diverse barriers. Further, the barriers are in Annex I countries as well as non–Annex I countries.

Paradigm II implies the need for more attention to South-North and South-South and even North-North technology flows, as well as the traditional focus on North-South flows in government policymaking and international negotiations. In this context, the role in international technology transfers of multinational firms based in developing countries—and barriers to them—need to be examined more thoroughly.

41. WTO (2008).
42. On international institutional architecture issues more generally, see especially Aldy and Stavins (2007).

This suggests at least one obvious basis for a *division of labor* in the emerging international institutional architecture for the joint climate-trade agenda: The UNFCCC, World Bank, and other international development institutions should continue to focus their efforts on North-South financial flows and capacity building in developing countries to enhance technology transfers. At the same time, the WTO and other trade-investment institutions should substantially increase their efforts to reduce barriers to trade and FDI in climate-related technologies. These efforts should include barriers in Annex I countries as well as non–Annex I countries. Further, the efforts should be undertaken with a recognition that barriers in non–Annex I countries are barriers to the exports and FDI from other non–Annex I countries as well as Annex I countries.

At the multilateral level, these efforts can be conducted largely on *parallel tracks*—one involving the international climate change and international development institutions, the other involving the international trade-investment institutions. However, there should be a more *formalized liaison* established between the two sets of institutions. This could be done through many institutional arrangements, for instance, a climate-trade liaison committee that includes the UNFCCC, United Nations Environment Program, WTO, and World Bank.

There is, however, a much more complex and challenging problem in the current array of *bilateral, regional, plurilateral*, and *multilateral* arrangements. The newly formed international institutional arrangements for climate change, in addition to the already-existing ones for trade and investment, have created a "system" that is doubly fragmented as far as the joint climate-trade agenda is concerned. Some of the bilateral, regional, and plurilateral arrangements might be justified on administrative and economic grounds as well as domestic and international political grounds. However, there are daunting analytic and policymaking challenges in trying to integrate them into the new international institutional architecture that will likely continue to have a wide array of multilateral climate, trade-investment, and financial institutions at their center.

There are yet more complex and significant issues about the relationships between technology agreements and other elements of the post-2012 climate and trade regimes. For instance, access to technology transfer funds or denial of access to them—including those already in existence and others yet to be created—could be used as "carrots" or "sticks" to create incentives for governments to undertake unilateral mitigation measures or to meet their internationally agreed-on obligations. Another example is the possible use of trade-investment-technology transfer sanctions to deter "free riders" on multilateral agreements.[43]

---

43. See Brewer (2003, 2004).

Finally, no matter how international technology transfer issues may evolve on the joint agenda for climate change and trade-investment-technology transfer, significant domestic political forces will shape the agenda and influence governments' and firms' choices. Those forces will reflect the familiar features of the domestic politics of international climate change and international trade-investment-technology transfer issues. Whether they will prevent progress in the development of constructive responses to the new joint agenda—and how to overcome them—is yet another issue for further analysis.

Comment

Comment by Muthukumara Mani

Although the largest share of historical and current global emissions of greenhouse gases has originated in developed countries, developing countries will soon account for a greater share in the growth of world carbon dioxide (CO_2) emissions from fossil fuel combustion than developed countries. The International Energy Agency's projections suggest that non–Annex I countries will overtake the Annex I countries as the leading contributors to global emissions in the 2030s. Non–Annex I countries' share of global emissions will soar from 38 percent in 2002 to 52 percent in 2030, while Annex I countries' share will decline from 60 to 47 percent. In other words, more than 70 percent of the world emissions increase from 2020–30 will come from non–Annex I countries. China alone will contribute about a quarter of the increase in CO_2 emissions, or 3.8 gigatons, reaching 7.1 gigatons in 2030. Strong economic growth and heavy dependence on coal in industry and power generation will contribute to this trend.

Countries (both developed and developing) have implemented various polices and measures to achieve their targets and have shown some progress in mitigating the effects of climate change. However, in a number of cases, economic considerations have far outweighed their consideration for the global climate. While a number of low carbon growth options exist for reducing our net greenhouse gases, the key to achieving this would be to move toward low carbon or no-greenhouse-gas-emitting technologies.

There are already a number of low-carbon technologies available to combat climate change, and it is now widely accepted that the stabilization of greenhouse gas concentrations to as low as 450 parts per million CO_2 equivalent can be achieved by deploying currently available technologies, which are expected to be commercialized in the coming decades. It could come in a most effective and efficient manner by allowing low-carbon growth in developing countries.

In this regard, international technology transfer can be a significant and cost-effective component of climate mitigation efforts in developing countries that balances their economic considerations. Current mechanisms of technology transfer through the Kyoto Protocol's Clean Development Mechanism have proved cumbersome and inadequate. Therefore, technology transfer and diffusion through the traditional channels of trade and investment could be important sources of climate change mitigation, as indicated by Thomas Brewer in chapter 4.

The Stern Review in fact identifies the transfer of energy efficient and low-carbon technologies to developing countries as critical to reducing the energy intensity of production. It further observes that "the reduction of tariff and non-tariff barriers for low-carbon goods and services, including within the Doha Development Round of international trade negotiations, could provide further opportunities to accelerate the diffusion of key technologies." However, in trade and investment regimes, one encounters a different set of issues and problems.

Recent efforts to liberalize trade in environmental goods under the World Trade Organization have had their share of problems. The fundamental fault lines of disagreement between countries are underpinned by different perceptions on what "environmental goods" are (that is, the issue of definition), which would determine what goods to include or not for liberalization under the mandate and how to liberalize (that is, the issue of approaches to liberalization) in a manner that addresses the interests of both developed and developing countries.

The key issues surrounding what to liberalize include (1) dealing with single versus dual-use goods; (2) the relative environmental friendliness of goods; (3) dealing with constantly evolving technology; (4) assessing implications for domestic industries, especially in developing countries; (5) dealing with non-tariff barriers; (6) enhancing opportunities for developing country exports; and (7) dealing with agricultural environmental issues.

Even if trade is or is not liberalized, foreign direct investment can be another important means of transferring technology. However, weak intellectual property rights (IPR) regimes (or perceived weak IPRs) and other barriers in developing countries often inhibit diffusion of these specific technologies beyond the project level. These barriers range from weak environmental regulations, fiscal feasibility, financial and credit policies, economic and regulatory reforms, and the viability of technology to local conditions (including availability of local skills and know-how). Streamlining of intellectual property rights, investment rules, and other domestic policies could further aid in widespread assimilation of clean technologies in developing countries

There is, therefore, a need in the absence of a cohesive global regime on technology transfer to better understand the policy and regulatory framework that would facilitate a greater deployment and diffusion of clean energy technologies for climate change mitigation through the broader trade and investment channels in both developed and developing countries.

References

Agra Europe. 2006. *FO Licht's World Ethanol & Biofuels Report.* London.

———. 2007. *FO Licht's World Ethanol & Biofuels Report.* London.

Alavi, Rokiah. 2007. "An Overview of Key Markets, Tariffs and Non-tariff Measures on Asian Exports of Select Environmental Goods." Issue Paper No. 4. Geneva: International Centre for Trade and Sustainable Development.

Aldy, Joseph E., and Robert Stavins, eds. 2007. *Architectures for Agreement: Addressing Global Climate Change in the Post-Kyoto World.* Cambridge University Press.

Anger, Niels, Chrisoph Bohringer, and Ulf Moslener. 2007. "Macroeconomic Impacts of the CDM: The Role of Investment Barriers and Regulations." *Climate Policy* 7: 500–517.

Barton, John H. 2007. "New Trends in Technology Transfer: Implications for National and International Policy." Issue Paper 18. Geneva: International Centre for Trade and Sustainable Development.

Brewer, Thomas L. 2003. "The Trade Regime and the Climate Regime: Institutional Evolution and Adaptation." *Climate Policy* 3: 329–41.

———. 2004. "The WTO and the Kyoto Protocol: Interaction Issues." *Climate Policy* 4: 3–12.

———. 2008a. "International Energy Technology Transfers for Climate Change Mitigation: What, Who, How, Why, When, Where, How Much, . . . and the Implications for International Institutional Architecture." Paper prepared for CESifo Venice Summer Institute Workshop: Europe and Global Environmental Issues, Venice, July 14–15. Available at www.usclimatechange.com.

———. 2008b. "The Technology Agenda for International Climate Change Policy: A Taxonomy for Structuring Analyses and Negotiations." In *Beyond Bali: Strategic Issues for the Post-2012 Climate Change Regime*, ed. Christian Egenhofer. Brussels: Centre for European Policy Studies.

———. 2008c. "Climate Change Technology Transfer: A New Paradigm and Policy Agenda." *Climate Policy* 8: 516–26.

Brewer, Thomas L., and Stephen Young. 2000. *The Multilateral Investment System and Multinational Enterprises.* Oxford University Press.

———. 2009. "The Multilateral Regime for FDI: Institutions and Their Implications for Business Strategy." In *The Oxford Handbook of International Business*, 2nd edition, ed. Alan M. Rugman. Oxford University Press.

Cambridge Energy Research Associates. 2008. *Crossing the Divide: The Future of Clean Energy.* Cambridge.

Cantwell, John. 2009. "Innovation and Information Technology in the MNE." In *The Oxford Handbook of International Business*, 2nd edition, ed. Alan M. Rugman and Alain Verbecke. Oxford University Press.

DeConnick, Helen, Frauke Haake, and Nico van der Linden. 2007. "Technology Transfer in the Clean Development Mechanism." *Climate Policy* 7: 444–56.

Egenhofer, Christian, Lew Milford, Noriko Fujiwara, Thomas L. Brewer, and Monica Alessi. 2007. *Low Carbon Technologies in the Post-Bali Period: Accelerating their Development and Deployment.* ECP Report 4. Brussels: European Climate Platform.

European Commission. 2003. *Report on Climate Change Technologies.* Brussels.

Fischer, Carolyn, Christian Egenhofer, and Monica Alessi. 2008. "The Critical Role of Technology for International Climate Change Policy." In *Beyond Bali: Strategic Issues for the Post-2012 Climate Change Regime,* ed. Christian Egenhofer. Brussels: Centre for European Policy Studies.

Gallagher, Kelly Sims. 2006. *China Shifts Gears: Automakers, Oil, Pollution, and Development.* MIT Press.

Howse, Robert, and Petrus B. van Bork. 2006. *Options for Liberalising Trade in Environmental Goods in the Doha Round.* ICTSD Series on Trade and Environment 2. Geneva: International Centre for Trade and Sustainable Development.

ICTSD (International Centre for Trade and Sustainable Development) ed. 2006. *Linking Trade, Climate Change and Energy.* Geneva: International Centre for Trade and Sustainable Development (http://sarpn.org.za/documents/d0002330/Trade_climate-change_ICTSD_ Nov2006.pdf [March 2009]).

IEA (International Energy Agency). 2008. *Energy Technology Perspectives 2008.* Paris.

IEA/OECD (International Energy Agency and Organization for Economic Cooperation and Development). 2006. *Energy Technology Perspectives: Strategies and Scenarios to 2050.* Paris.

Lehman Brothers. 2007. *The Business of Climate Change, I and II.* New York.

Lewis, Joanna I. 2007. "Technology Acquisition and Innovation in the Developing World: Wind Turbine Development in China and India." *Studies in Comparative International Development* 42: 3–4.

OECD (Organisation for Economic Cooperation and Development). 2001. *Environmental Goods and Services: The Benefits of Further Global Trade Liberalisation.* Paris: OECD.

Pacala, Stephen W., and Robert H. Socolow. 2004. "Stabilization Wedges: Solving the Climate Problem for the Next 50 Years with Current Technologies." *Science* 305: 968–72.

Sell, Malena (2006). "Trade Climate Change and the Transition to a Sustainable Energy Future: Framing the Debate." In ITCSD, *Linking Trade, Climate Change and Energy.*

Socolow, Robert H. 2006. "Climate Change: Princeton Professor Lays Out Broad Strategy on Greenhouse Emissions." Presentation at World Bank's 2006 Energy Week, Washington, March 9.

Socolow, Robert H., and Stephen W. Pacala. 2006. "A Plan to Keep Carbon in Check." *Scientific American,* September, 50–57.

South Africa. 2006. "Ministerial Indaba on Climate Action." Kapama Lodge, South Africa, June 17–21.

Steenblik, Ronald. 2006. "Liberalisation of Trade in Renewable Energy and Associated Technologies: Biodiesel, Solar Thermal and Geothermal Energy." Trade and Environment Working Paper No. 2006-01. OECD.

Stern, Nicholas, 2007. *The Economics of Climate Change: The Stern Review.* Cambridge University Press.

Sugathan, Mahesh. 2006. "Climate Change Benefits from Liberalisation of Environmental Goods and Services." In ITCSD, *Linking Trade, Climate Change and Energy.*

UNEP (United Nations Environment Programme). 2007. *Global Trends in Sustainable Energy Investment, 2007.* UNEP.

UNFCCC (United Nations Framework Convention on Climate Change). 2002. "Expert Group on Technology Transfer: Programme of Work, 2002–2003." Bonn: UNFCCC.

———. 2004. "The First Ten Years." Bonn: UNFCCC.

———. 2006a. "Recommendations of the Expert Group on Technology Transfer for enhancing the implementation of the framework for meaningful and effective actions to enhance the implementation of Article 4, paragraph 5, of the Convention: Note by the Chair of the Expert Group on Technology Transfer." Bonn: UNFCCC.

———. 2006b. "Synthesis report on technology needs identified by Parties not included in Annex I to the Convention." Bonn: UNFCCC.

———. 2006c. *Technologies for Adaptation to Climate Change.* Bonn: UNFCCC.

———. 2007a. "Work Programme of, the Expert Group on Technology Transfer (EGTT) for 2007." Bonn: UNFCCC.

———. 2007b. *Investment and Financial Flows to Address Climate Change.* Bonn: UNFCCC.

U.S. Climate Change Technology Program. 2006. "U.S. Climate Change Technology Program Strategic Plan" (www.climatetechnology.gov/library/2006 [September 2006]).

U.S. Office of the Trade Representative (USTR), 2006. *Report by the Office of the United States Trade Representative on Trade-Related Barriers to the Export of Greenhouse Gas Intensity Reducing Technologies.* Washington.

Vattenfall. 2007. Climate Map (www.vattenfall.com [February 2009]).

World Bank. 2006a. *Clean Energy and Development: Towards an Investment Framework.* Development Committee of World Bank and the International Monetary Fund. Washington.

———. 2006b. *An Investment Framework for Clean Energy and Development, Progress Report.* Vice President for Sustainable Development. Washington.

———. 2007. *An Investment Framework for Clean Energy and Development.* Washington.

———. 2008. *Warming Up to Trade: Harnessing International Trade to Support Climate Change Objectives.* Washington.

WTO (World Trade Organization). 2008. Regional Trade Agreements Chart (www.wto.org [March 2008]).

WILLIAM ANTHOLIS 5

Five "Gs": Lessons from World Trade for Governing Global Climate Change

Reversing the greenhouse gas (GHG) emissions of the world's $60 trillion economy will be among the most complex international governance challenges ever—rivaling the forty-year effort to dramatically reduce tariffs and establish a rules-based trading system. Given that nearly fifteen years have passed since the completion of the last global trade pact, it is easy to forget that the World Trade Organization (WTO) stands tall among the great successes of global governance, precisely because it was so difficult to accomplish. A counterpart twin tower—a global system to address climate change—can mimic the trade regime's most successful governance principles, and learn from its structural weaknesses. Perhaps more important, as this volume's theme suggests, the two regimes need to work diligently to avoid colliding with one another. Indeed, it would be both unfortunate and ironic if a global climate regime only could succeed at the expense of the global trade regime—or vice versa.

What lessons should the climate regime learn from the trade regime? It may be helpful to break the issue down into five core questions for any attempt to govern: Who governs? What is the structure of the basic governing agreement? Where is it "binding"? When can we expect the agreement to take effect? How does it bring new nations in? For each question, preliminary answers can be found in what we might think of as the five "Gs" that should govern climate change. By looking to the lessons from the WTO, I try to make the case for a climate regime that:

—starts with a *group* of major emitters, which together
—forge a *general agreement* to tackle the issue, one that
—*gears up* nations' domestic action and that
—organizes itself around a *generational* goal that
—allows for the *graduation* of developing countries into full commitments.

The author is indebted to comments from Scott Barrett, Colin Bradford, Lael Brainard, Daniel Drezner, Stuart Eizenstat, Alex Fifer, Lauren Fine, Warwick McKibbin, Carlos Pascual, Nigel Purvis, David Sandalow, and Strobe Talbott.

In a few of these areas, such an approach can provide a road map to resolving potential conflicts between the two regimes.

Who Governs?
The Right *Group* of Nations, Matched to the Challenge

International regimes need to be designed to their purposes. Are they debating forums? Are they negotiated agreements that govern in particular fields? Trade and climate change have both benefited considerably from both kinds of organizations. This chapter assumes that concerned nations are moving toward a governing regime for GHG emissions, and that they need mechanisms equipped to address that challenge.

Since the formation of the United Nations system, two bodies have existed along side one another on the issue of global trade, one for discourse, the other for governance. The UN Conference on Trade and Development (UNCTAD) has largely functioned as a forum for assessing the twin goals and accomplishments of trade and development. Alongside it, the General Agreement on Tariffs and Trade (GATT) and its successor, the WTO, have been the governing body for global trade. Though some might find it odd to point to the WTO as a successful model of international governance (especially given recent difficulties in completing the Doha Round of multilateral trade negotiations), it is easy to forget how significant its contributions have been to both international cooperation and to economic growth over the last sixty years.[1] The GATT/WTO system began as both a smaller (in terms of membership) and more ambitious (in terms of governance) world body than UNCTAD when a group of the right countries decided to work together.

Lesson learned: Size matters. When it comes to global governance, it was and is easier to get things done with a smaller number of the right countries. The GATT process was managed by the biggest and most technically competent trade players—the so-called Quad of the United States, Japan, Canada, and Europe. Occasionally, when formal negotiations bogged down, the Group of Seven (and later, the Group of Eight) would weigh in to give the talks a boost, such as in 1978 and 2001, when the leaders themselves helped spur break-

---

1. That success was apparent twenty-five years ago, when the GATT system was held up as the model for global governance—including among "realist" theorists of international relations, who tend to hold a dim view for institutions. See Ruggie (1983). Though Ruggie would not classify himself as a realist, his general argument was accepted by realists such as Krasner. In the real world of politics, the GATT and WTO's acceptance among American political conservatives— including their willingness to accept binding decisions by international tribunals—is striking.

throughs leading, respectively, to the close of the Tokyo Round and the launch of the Doha Round.

As the WTO's membership grew in size over its first five decades, negotiations became more unwieldy. The greatest number of new entrants came from developing countries. After an initial sorting out, the lesson of size was relearned: A new Quad was established, where India and Brazil joined the United States and European Union as the principal negotiators.

Further complicating matters, over the years, a plethora of regional and bilateral agreements have advanced trade liberalization worldwide. The EU has led the pack in depth of integration and effectiveness, but the last forty years have seen the rise of a South American commercial union (Mercosur), the North American Free Trade Agreement, the Southern African Customs Union, and the Association of Southeast Asian Nations' Free Trade Area. Of course, there is considerable debate about whether this spaghetti bowl of different agreements has been good for the global trading system. Supporters of the three way street (that is, global, regional, and bilateral) have found "competitive liberalization" to be a positive force. Regional agreements help drive reluctant countries to the global negotiations for fear of missing gains from trade. Opponents see the growing complexity and difficulty of multiple trade talks to exceed the negotiating capacity of diplomats and the political will of elected officials. Complexity is unavoidable, to be sure. That the complexity has been at all manageable is due, in part, to the bedrock of a rules-based system that was established sixty years ago, and the committed leadership of a relatively small number of players.

So what does this mean for the climate change regime? The half-true cliché about climate change is that it is a global problem that requires a global solution. Still, moving forward does not require all countries to be part of the solution—at least not at first. The UN-sponsored Kyoto Protocol process was slowed down by trying to conduct a global research initiative on the nature of the challenge (largely led by the UN's Intergovernmental Panel on Climate Change), while also debating who was responsible for addressing the challenge and negotiating an agreement among 140 nations under the UN's Framework Convention on Climate Change. Though data, debate, and dialogue were critical to convincing these nations of the challenge at hand, the negotiations over what to do about it became rancorous and left many questions unanswered. They gave way to several more years of disputed talks on how to implement the agreement, a lengthy and unsuccessful ratification discussion in the United States, and uninspiring results on the ground—even from enthusiastic backers like the European Union and Japan, which face an uphill battle in meeting their 2008–12 GHG emission targets. Meanwhile, the main developing country bloc is an eclectic group, including nations ranging from giant powerhouses such as Brazil, China, and India to small, poor, landlocked nations in Africa to small island nations. With the exception of these island countries—which literally could get washed away if there is no progress—most have been quite comfortable with the UN's penchant for discussion, so long as those discussions do not lead to binding obligations for their own economies.

In short, we have a potentially large problem coupled with a complicated, bureaucratic, and torpid negotiating mechanism. If size matters when setting up a governing regime, then the climate system needs to separate the broad and inclusive dialogue about the challenge from the more narrow and detailed challenge of negotiating an agreement. The latter task is best taken by a smaller group of nations.[2]

The great bulk of GHG emissions likely to spew into the atmosphere over the next three decades—not to mention the economic and technical capacity to reverse course—can be found in fewer than two dozen countries. The creation of smaller groupings—such as a Major Emitters' Group (E-8)—could help to address these challenges.[3] The United States, European Union, China, Russia, Japan, and India are the top six emitters of GHGs; and South Africa and Brazil rank tenth and thirteenth, respectively, but their contributions are significant in representing their regions—especially Brazil, where protecting the Amazon River Basin is a major priority in storing carbon. This same logic lies behind the major emitters' meeting that President George W. Bush hosted in September 2007, which adds to my list of eight and includes Canada (seventh), South Korea (eighth), Mexico (ninth), Indonesia (twelfth), and Australia (fifteenth). Together, these thirteen countries produce more than 80 percent of all GHGs.

Keeping the core group of negotiating nations small—and, occasionally, involving heads of state in the conversations—has one other signal virtue. The same set of players is at the center of WTO negotiations. As the two regimes begin to bump into each other on a range of issues—from border surcharges

2. One commentator questioned whether the problem of protecting the Earth's climate is analogous to that of expanding free trade. As a general matter, most analysts would agree that protecting the climate is a nonexcludable public good, whereas free trade has been less so, since only the participants in a trading regime enjoy the benefits. Some might even question whether free trade is a public good, though cf. Kindleberger (1986). Indeed, a strong argument can be made that both a climate regime and a trade regime offer both excludable and nonexcludable public goods. In trade, the excludable public goods are the lower tariffs and trade barriers offered to the members of the regime; the nonexcludable good is the stable international economic order that has economic and political benefits for all countries. In climate change, the nonexcludable good would be climate protection. The excludable good would be an emissions trading regime. Many thanks to Lael Brainard for helping clarify this distinction.

3. Stern and Antholis (2007b).

to energy subsidies—resolution can be reached more easily if the same players from both regimes are talking. That is especially true if heads of state themselves are aware of the need to coordinate, and the perils of the failure to do so.

What Is the Form of Governance? A *General* Agreement

One of the keys to the GATT/WTO's success is that it did not start as a global body but rather as a less formal arrangement. If this distinction seems unimportant, keep in mind that the WTO started not as the successful WTO, or even the successful GATT, but as the failed International Trade Organization—which was envisioned at Bretton Woods, along with the World Bank and International Monetary Fund, and whose treaty died on the U.S. Senate floor, because two-thirds of that august body was not prepared to hand over highly political decisions regarding trade policy to an international organization. The negotiators went back to the drawing board. Only after the International Trade Organization's high-profile failure did they come up with the *General* Agreement on Tariffs and Trade.

The core lesson: Do not start with an international treaty organization responsible for data, debate, and enforcing compliance. And when it comes to enforcement, build confidence through general agreements, which are "binding" in that they synchronize and increase the ambition of domestic action that states see as being in their best interest. For nearly fifty years, the GATT was able to negotiate and adjudicate agreements that bound nations in a way that less directly called national sovereignty into question. Each participating nation pledged to cut tariffs and other trade barriers in a coordinated way. Countries could choose what counted as significant reductions, and they would often trade fast action in one area for slow action in another. Once commitments were made, they had to be enforced. An adjudicative body was established to resolve trade disputes.

Technically speaking, the adjudicative trade body did not enforce the treaty. Member nations did. Countries monitored one another's behavior—including the most economically powerful trading nations. When a plaintiff country had a complaint, it brought it to the GATT/WTO's dispute resolution mechanism. If a defendant country lost a dispute, it had a choice: Change its domestic law, or allow a retaliatory tariff or other action by the plaintiff country. In this way, all countries felt the system to be self-enforcing. All this gave negotiators the ability to say convincingly to their political masters—including general publics—that the agreement was not a sacrifice of sovereignty.

The fear that nations will lose their sovereignty similarly has plagued the climate change discussions. If the United States had ratified the Kyoto Protocol, it would have been a "binding treaty." Opponents of Kyoto claimed that the United States would have been liable for some set of sanctions that would be administered and enforced by the mandates of the United Nations. America's sovereignty over its energy future—and by extension, its national security—would be subject to external intervention. As a political matter, few American politicians want to be told that they must do something, or else face sanction by a global body.

Whether or not those concerns have any factual merit, "sovereignty hawk" nations around the world (particularly the United States and much of the developing world) have feared Kyoto-style obligations. Political leaders in the United States, China, India, and Brazil also have refused to sacrifice their ability to control their economic destinies to a global energy regime—at least, not give up that sovereignty in a way that diverges from national interest. Only the European Union—whose members have grown comfortable sharing or even pooling their sovereignty—seems to like the idea of using an international agreement to compel domestic action.

There is another way, of course. Building on the successful GATT model, negotiators could seek a General Agreement to Reduce Emissions (GARE). Like the GATT, the GARE would effectively link domestic action with an international agreement.[4] It would avoid moving too quickly to a full-blown international institution, such as a World Environment Organization. If a "treaty" suggests that nations are tying their fates to one another, "general agreements" suggests that nations acknowledge one another's interdependence, but also their autonomy. As they build confidence in their ability to work together under such agreements, they may become more willing to strengthen the regime.

A GARE system could be built on the E-8 or major emitters' group outlined above. A core set of the most important countries could start the process, and this ultimately would be compatible with regional and bilateral agreements. On an annual basis, leaders of this group could meet at the summit level to evaluate progress and to help give a boost to the ongoing negotiations.

What then of the United Nations? An important role remains for the UN in continuing to sponsor the broader climate talks as a forum for helping nations share information and best practices with one another. The UN also has been pathbreaking in supporting the critical work of the Intergovernmental Panel on Climate Change—the scientific body that has helped establish that climate change is real, and that human action is contributing dramatically. Both these

4. See the first suggestion for such an approach in Stern and Antholis (2007a).

functions help support the negotiation and conflict resolution functions of a binding agreement. Eventually, once confidence is built in a self-enforcing agreement, the UN can be brought in to maintain the relationships.

Where Does the New Climate Regime Bind Nations?
It *Gears Up* Domestic Steps Nations Are Willing to Take

Ask a State Department lawyer, and she will tell you that there is no difference between a treaty, a congressional-executive agreement, and a presidential bilateral statement with a foreign head of state. The United States is honor-bound to live up to its agreements, whatever form they take. If the agreement includes consequences for violation, the United States is obligated to accept those. Yet in practice, nations (including the United States) frequently violate or ignore agreements—and either suffer the consequences or do not. Though the UN Charter provides some instances when states may be physically compelled to act in accord with violating international norms, in practice this rarely is the case for nonmilitary agreements.

What makes some international agreements binding? What makes some bindings succeed and others fail? There are at least three ways to discuss the success of binding agreements. First, some pacts succeed because states feel no need to violate them. These agreements succeed because they create a structure that allows states to do what they would prefer to do, but might not do because they fear noncompliance by others. By giving states confidence that other states will live up to their end of the bargain, agreements allow states to do what is in their best interest. This is what de Tocqueville called "self-interest rightly understood."

Second, some agreements succeed because nations realize, upon violating an agreement, that the net costs of doing so are not worth it. This is usually the case when nations contemplate sanctions from an agreement—and the political impact those sanctions could have domestically and internationally—and choose to get right by the law. Third and last, agreements work when nations suffer appropriate consequences for their violations, and both the violating nation and the nation that applies the sanction feel the consequences to be appropriate and adequate.

In theory, all three cases do not require an outside enforcing body. It is governance without government, or what the great international relations theorist Hedley Bull called "the efficacy of international law," which "depend[s] on measures of self-help."[5] The GATT/WTO succeeded because, for its first fifty

5. See Bull (1977, 131, and chapter 6).

years, all three forms of self-help worked. First, the commitments were suffi-ciently robust that countries could plan to cut trade barriers—that is, gear up their commitment—knowing that counterpart nations would do the same. GATT/WTO negotiations helped nations to cut their own trade barriers further than they otherwise would. In return, counterpart nations also lowered their barriers. Consumers benefited from cheaper imports, and exporters benefited from wider markets. Nations understood the tough domestic challenges other nations felt in trying to lower trade barriers.

This worked in practice, particularly when Congress signaled its willing-ness to lower barriers in specific product areas in advance of a negotiation. Making a priority of domestic action is actually enshrined in the domestic legal architecture of American trade diplomacy. From an American perspective, one reason that the United States is more easily bound by trade negotiations is that it uses congressional-executive agreements, which require passing relatively detailed trade promotion authority in advance of negotiations. As a result, the trading system aspired toward laissez-faire goals as a general matter across national boundaries, but also accepted that national legislation was central to moving forward. Though laissez-faire remained a long-term goal, no single round or negotiation ever proposed to complete the process and each succes-sive round depended on national action. The system recognized the domestic political and economic constraints that nations face in moving toward a glob-ally integrated goal.[6]

Second, the GATT's enforcement system sustained national cuts without appearing to undermine sovereignty. When a nation was found to be in viola-tion of a trade rule, it had a choice: Change its trade practice, or accept reciprocal trade sanctions on other goods. Even under trying circumstances, nations were willing to go back and change domestic law in order to come into compliance. In these instances, countries have avoided the imposition of sanctions, and they have been unwilling to sustain extended tit-for-tat sanctions. Third, in those few cases where sanctions have been applied, nations have generally been will-ing to accept them without countersanctions. Rather than starting trade wars, the GATT/WTO system has prevented them.

A similar logic can guide a GARE: Countries can choose domestically to cut their GHG emissions in the way that makes most sense, given their domes-tic constraints. Rather than prioritize a "treaty" as a goal in and of itself, a GARE would start with domestic legislation and help nations strengthen—that is, gear up—their ambition.

Nearly all nations recognize that cleaner energy production and the protec-tion of forests are worthwhile goals in themselves, and that they should act to

6. See Ruggie (1983).

prevent irreversible climate change. Almost all nations have taken some steps in this regard. And a diversity of approaches is appropriate. Countries use energy and regulate pollution very differently, and they also differ widely in their capacity to track emissions and enforce compliance. The United States and China, for instance, are particularly dependent on carbon-intensive industries such as coal. Brazil, conversely, has huge sources of renewable resources, such as hydropower and biofuels, but is also struggling to save its rain forest—one of the great carbon reserves and "sinks" in the world. It is clear that a one-size-fits-all approach will not work.

The threefold challenge for the international negotiations is, first, how to get countries to take reciprocal domestic actions; second, how to structure compliance so that it reinforces or returns states to mutual action; and third, how to establish sanctions that nations can choose to accept as appropriate. Thus, first, a GARE should begin in domestic action, and use the negotiating process to gear up the ambition of states. States are "bound" to follow through on actions they are likely to take on their own.[7] One way to make sure that that is the case is to legislate first and negotiate later. In the American context, GARE would take advantage of congressional-executive agreements and avoid the treaty process. In a GARE, the domestic political hurdle to passage is whether to pass and implement a domestic law. With the framework of such a domestic law in place, the international negotiations can focus on the level of ambition that all countries take, so as to help ratchet up ambition. The diplomatic challenge becomes whether that level of commitment is acceptable to counterpart nations.[8]

This is in slight, but significant, contrast to the Kyoto Protocol's approach of binding a state to an international organization's decisionmaking.[9] For instance, in the United States, the treaty process not only requires the supermajority in one house of Congress; it also requires passage of implementing legislation in both houses. Agreements, by contrast, require majorities in both houses—first for authorization to negotiate, second for the final agreement itself.

7. For an overview of what a domestic and international approach for the United States might look like, see Stern and Antholis (2007a).

8. One model example for this would be the EU's proposal to unilaterally cut their emissions by 20 percent below 1990 levels in the post-Kyoto commitment period, and to extend those cuts to 30 percent if an international agreement is reached.

9. One advantage by not being a treaty, the GARE would avoid another major drawback of Kyoto: It would not need a two-thirds majority in the United States Senate, a minefield where countless treaties have gone to die. Indeed, by the treaty process, internationally agreed emissions targets and timetables the policies and regulations needed to comply with them become deeply enshrined in domestic law as they have been passed by a supermajority in the Senate. By contrast, the GARE would only require simple majorities in both the House and the Senate, putting the domestic legislation horse in front of the global treaty cart—just the way it should be. See Stern and Antholis (2007a); Purvis (n.d.).

The authorization to negotiate—so-called fast track in trade talks—gives negotiators a road map for what can be negotiated, and as a result begins to involve
members of Congress in the talks themselves.[10] In a real sense, for the United
States, a GARE would start with domestic action, and seek to ratchet it upward,
in sync with other nations.

Second, a GARE would need to be "binding" by addressing noncompliance.
As with the early GATT system, it should include avenues for self-enforced
sanctions by nations. Exactly how nations will self-enforce an agreement is
still being debated. Some analysts have called for a common global carbon tax.
Others have called for a "pledge and review" process, in which nations pledge
to reduce GHGs, and then review one another's progress on a regular basis.
There may be merit to both kinds of agreements. Yet neither one, on its face,
appears to encourage the gearing up of domestic commitments, while also dis-
couraging nations from breaking those commitments by imposing sanctions
that deny nations the benefits of the agreement.[11]

One approach, in theory, does accomplish these goals: international trading
of GHG emissions. As a domestic matter, the EU has already adopted emissions trading, and the United States is considering such legislation, having
successfully pioneered a sulfur dioxide system under George H. W. Bush in
the late 1980s. Though there have been some initial problems with the EU's
system, it has now done largely what it intended to do: put a price on emissions, and create incentives for the private sector to find emissions cuts where
most efficient to do so.

International emissions trading would extend these advantages across
national borders. The United States insisted on GHG emissions trading at
Kyoto, and for nearly two years afterward haggled with the European Union
over the rules. Ironically enough, once the United States walked away from
emissions trading during the George W. Bush presidency, the European Union
began to aggressively pursue international emissions trading. Trading can happen in two forms—in either a closed or an open system. In a closed system,
two different national economies agree that total emissions in both economies
will not exceed a fixed amount. As long as both nations comply in the aggregate, permits remain of equal value and are freely tradable between countries.
If one country violates its emissions limits, however, the permits in that country become less valuable. In an open system, nations are responsible only for
their own reductions, though investors or companies may seek certifiable reductions in other countries, and simply be free to invest in such reductions.[12]

10. See Antholis and Talbott (2007).
11. For a useful discussion on this, see Wiener (2007, 74–76).
12. As mentioned above, establishing an emissions trading system would move from the nonex-

Both approaches have strengths and weaknesses from a "compliance as self-help" standpoint. The strength of the closed system is that it raises the stakes for compliance—and the penalties for noncompliance. In such a system, it is highly advantageous for nations to make broad progress on their GARE reduction commitments, as it would either force nations to seek permits from firms that have successfully cut GHG emissions in other nations, or provide incentives for nations to have the most number of such firms in their own territory. If it were possible to set up such a system, the incentives for success should be high. Yet the cost of failure should also be high, as less successful countries would be forced to pay dearly for emissions permits across borders. In contrast, an open system would create incentives for investing across borders. That said, it would provide few downsides if nations failed to comply with the international agreement—other than the greater risk of failing to stabilize the climate.

The joint challenges for a GARE that relied on trading for compliance would be to determine whether a member country seeking to join had proposed a strong enough target, and whether preexisting members had come close enough to their previous commitments in each successive round of negotiations. The first task would need to fall to member states. The second task could fall to a joint review panel established by GARE countries. If a country failed to meet its target by reducing its emissions or buying permits, it would forfeit the right to continue in the GARE in future periods.[13]

Third, establishing a successful binding agreement requires addressing how to deal with those who refuse to join. A growing chorus is raising the idea of using actual trade protections—such as demanding that imported goods from countries that have not adopted sufficient emissions reductions would need to purchase emissions permits equivalent to their carbon footprints. The idea first arose in countries such as France, directed at the United States for not joining the Kyoto Protocol. Now that the United States is contemplating joining a post–Kyoto system, Americans are considering applying the same approach to developing countries that do not take binding targets. These "border permits" would be a way of placing some sanction on nations that refuse to join or comply with an emissions agreement—and thereby help share the cost of compliance.

This has the potential both to be a constructive way to think through the problem but also to undermine the trade regime, the climate regime, or both. The constructive element of such an approach would be to provide real lever-

cludable public good system of climate protection to a system with excludable benefits: access to trading with other parties, with the enhanced efficiency and reduced compliance costs this implies. See n. 3, above.

13. See Stern and Antholis (2007a, 183).

age for nations to actually transfer the costs of noncompliance in an effort to address a global public good—something for which the trade regime allows exemptions.

The potential disruptive element is that all nations do not recognize the public good in the same way, let alone the means to address it. Developing countries, which likely would be the targets of such a system, are almost certain to claim that (1) this is a violation of the WTO's rules against nondiscrimination, and (2) that it does not meet the standard for environmental exemption for those rules. The "global public good standard," developing countries would likely argue, is not met because the current international climate treaty already embodies how the international community defines the climate challenge. That treaty, they will claim, explicitly demands that industrial nations act first, and that developing countries are exempt from binding targets. Because the standing global consensus is that industrial nations must act first, any effort to use the trade regime to shift that burden would be seen as illegitimate.

So if industrial countries persist in imposing such tariffs in order to build a more effective climate regime, they might undermine the WTO—regardless of which way the dispute settlement system determines the merits of the case. If a developing country claimed that this was a violation of WTO rules but lost the dispute, the victory for industrial countries would come as an additional blow to developing nations, on the heels of the WTO's long-stalled Doha development round, which has failed to produce market openings to industrial markets. Conversely, a victory for developing countries might further undermine public support for the WTO within industrial nations—which continues to wane. Likewise, the effect on the climate regime could be enervating. Emerging market players such as Brazil, China, and India will feel that they are being forced into a climate agreement by being denied access to an international trading regime that they have worked hard to enter as full participants. And industrial countries might be less inclined to join the climate regime if border adjustments are found to be illegal vis-à-vis the WTO, because they will feel their competitiveness further eroded.

Avoiding this clash of global governance regimes should be a priority for not only leading nations but also for the heads of both global regimes. It is perhaps the best argument for the world's leading economies to not treat these issues in isolation from one another, or from broader global economic developments. Indeed, one of the ironies of the spread of democracy has been that those governments have to work so hard to accomplish domestic regulation and, as a result, are often the least inclined to take direction from international organizations. The relatively fragile support for international regimes should not be easily challenged—particularly in the name of establishing other regimes.

When Can We Expect the New Climate Regime to Take Effect? Over a *Generation*

The idea of extending the enforcement of commitments over time gets at a central element of any governance challenge. One of the great successes of the trade regime was that it built itself gradually. Only after forty-five years of operating did it lead to a treaty organization.

The long-term nature of the climate challenge means that solutions must also be long term. Today's warmer climate is the result of GHG emissions accumulated over the last half century. Today's emissions add to those historic concentrations, and are already locking in warmer temperatures well past the middle of this century. Little can be done now to stop that warming from happening. So the effort to slow emissions over the next several decades will most affect temperature in the second half of this century.

What is the appropriate long-term goal? The starting point for all climate negotiations, the 1992 Rio de Janeiro Treaty (ratified by the U.S. Senate, and adopted worldwide), included an abstract long-term goal: "stabilization of greenhouse gas concentrations in the atmosphere at a level that would prevent dangerous anthropogenic interference with the climate system." The Kyoto Protocol was a practical attempt to implement Rio, yet it only set one target—a short-term reduction of GHG emissions by industrial nations. This was seen as a first step toward the longer-term goal. But because it lacked any second or third step, it was widely criticized for not getting at the longer-term challenges.

As with the trade regime, the climate regime should keep this long-term focus that was part of Rio's plan and be geared around a portfolio of long-term targets—including concentration levels and global temperature change. As with any law or diplomatic agreement, those targets could be adjusted later as scientific and economic evidence is collected. But the key is to get some agreement on the long-term goals so that short-term steps can be seen in their broader context.

Right now, many scientists believe that dangerous interference with the climate could be avoided if temperature increase is limited to 2°C. Consensus estimates predict that doing so requires at least stabilizing GHG concentration levels at 550 parts per million by 2050.

If the E-8 or a major emitters' group adopted 2°C and 550 parts per million as global goals—and urged other nations to do the same—countries could then target their short-term and long-term emissions cuts at levels that they felt to be effective and fair steps toward that goal. When diplomats try to negotiate over relatively short-term emissions cuts, they would be better able to explain to their political leaders and publics how each short-term step contributes to a longer-term effort. (Indeed, in the recent proposed Lieberman-Warner climate legislation, a series of emissions cuts are written in, extending out to 2050.) As nations reach their shorter-term benchmarks, they could assess how they are doing toward that longer-term goal. Among other things, this will help industrial countries signal to developing countries what they consider to be fair burden sharing for all nations over a future term, and that it is possible to achieve these marks without hurting economic growth.

Setting targets for temperature increase and gas concentrations can also help politicians, the media, and the public stay focused on the purpose of the undertaking: whether emission cuts are sufficient to slow and eventually stop global warming. Though scientists now overwhelmingly agree that human activities are leading to global warming, new evidence is coming in constantly. The consensus is being affirmed, but also challenged and updated on a nearly daily basis—mostly in the direction of sending more dire warning signals. Some scientists, for instance, now think that stabilization at 450 parts per million is needed to prevent 2°C of warming. Of greater concern, 2°C of warming may not be so safe. Recent research, for instance, finds that the current level of warming is melting the Arctic ice cap faster than had been anticipated, potentially weakening the ice cap's ability to reflect sunlight and cool the planet. If the ice cap were to disappear with less than 2°C of warming, it could be a tipping point that would lead to a more dramatic and dangerous shock to the Earth's climate.

How Does the New Climate Regime Bring New Nations into the Agreement? It Must Provide a Path for *Graduation*

Perhaps the greatest lesson the climate regime can learn from the trade regime is something that the latter has failed, so far, to entirely address: how to bring the developing countries into the regime in a way that acknowledges their development challenge but also allows them to graduate to full responsibility as their economies grow. The trading regime is now in the midst of its longest negotiating round in its sixty-year history—the so-called WTO Doha development round. One of the main reasons why it has been so difficult to conclude this round is that it is trying to address the regime's core weakness: that the two basic groups of main players—the industrial countries and the developing countries—have differing sets of obligations. The developing countries enjoy "special and differential treatment," which means that they are exempt from the more drastic tariff reductions taken by the industrial nations. Not only is the regime asymmetrical, but it is also unclear how any developing nation would graduate to taking on an industrial-strength obligation, when the time was right. Thus, although the addition of these developing countries has been critical to achieving global scope for the WTO, it also has added to the complexity of the process—and the current stalemate in negotiations.

As with the global trading system, the developing countries will ultimately need to graduate and become part of the post–Kyoto Protocol climate system. Kyoto was problematic in several ways, but perhaps its biggest drawback was that the developing countries did not commit to cut their GHG emissions—in fact, the treaty actually prevents them from taking a binding target even if they want to do so. Argentina, for instance, tried to take on a binding target in 1998, but it was prevented from doing so by other developing countries.

It certainly makes sense for the developing countries to have different obligations, or obligations that kick in later, given the industrial world's historic responsibility and much greater wealth, along with the generational nature of the problem. But there is simply no way to solve the climate problem without the active involvement of the developing countries—which, according to current projections, will account for more than 70 percent of GHG growth in next twenty-five years. Yet these countries show no willingness to accept Kyoto-style targets.

This catch-22 is not just a political problem; it is an economic one that goes to the heart of getting clean energy markets up and running. Most industrial countries are now poised to take near-term and middle-term efforts to cut GHG emissions, which is already leading to some increased investment in clean energy. However, if the world economy is going to cut its carbon emissions by as much as 80 percent, enormous amounts of capital investment will be required to find transformative, carbon-free sources of energy. The more certain investors feel that the industrial countries will keep seeking ever deeper reductions in emissions, the more likely they will be to commit that kind of capital up front. The key is for the industrial countries to signal their long-term cuts. But they are less likely to do so long as developing country action is not a sure thing. Right now, the developing countries are saying that they will not act, and they are refusing to address the long-term challenge.

How can the international community break out of this box? The effort must begin with the industrial world, by responding realistically to developing country concerns about equity. The developing countries rightfully feel that the rich countries are largely responsible for the problem to date, and probably for the global warming that will take place over the next fifty years. The industrial countries should not dismiss these concerns, particularly because the developing countries, particularly China and India, despite their recent economic gains, still have a nearly unfathomable number of their citizens living in extreme poverty—well over a billion people combined in those two countries alone. In addition to taking seriously efforts to estimate how much the industrial countries have contributed to current GHG concentration levels, these nations should also consider very-long-term targets on a per capita basis.

Second, the industrial countries should appeal to the developing countries' own self-interest. Climate change is most likely to hurt poor countries the worst, accentuating droughts and severe storms, for which these nations are least prepared. Moreover, many of these countries are facing the local air pollution that comes in the early stages of industrialization, and the health care challenges of clean air and water that could be lessened by early adoption of clean energy technology. Moreover, investing in energy efficiency and clean energy is ultimately cost-effective.

One possible motive for joining a GARE would be the potential to earn emissions trading credits on a sizable scale. In the near term, this would mean continuing to explore opportunities to earn emissions reduction credits on a project-by-project basis. This could potentially build support within the developing countries for adopting country-wide emissions policies linked to the GARE.

And last, the industrial countries should not be shy about public diplomacy on climate change. Right now, the developing countries do not feel any public pressure to respond to climate change—which is probably not surprising, given the development challenges many of them are facing. Thus, a public diplomacy strategy is needed that stresses each topic noted above—from equity to self-interest to the power of global markets to help transfer technology and capital to developing countries. Of course, all these efforts require that the real first steps be taken in the industrial world.

Conclusion

The political will has begun to develop in the United States and even in a few key developing countries for a global effort to reduce GHG emissions. This public support, however, still remains far from the dramatic shift in consensus needed to establish a full-blown global institution to address the climate challenge. In addition to the costs associated with acting, a core concern is a familiar one in global governance: loss of sovereignty. There is some good reason for this. Even for the most committed nations, the climate change challenge is of such great economic and environmental complexity that few politicians are likely to simply turn over the keys of their national policymaking to an international treaty organization.

In taking the first steps toward a global climate regime, the industrial nations can learn from the experience of how the global trading regime built confidence in a self-regulating system. The GATT/WTO system built on a small group of states that, through a general agreement, were able to gear up domestic action over a generation. The advantages of this approach are that it does not pose a direct challenge to national sovereignty. Instead, it coordinates the work of states in a way that respects the diversity of local governance, and has a greater chance of getting buy-in from the key players. The challenge of such an approach is that it does not guarantee fast domestic action, that many smaller states will feel left out of the process, and that the transition to the system may be difficult for many participating states. Last, as with the trade regime, it must overcome the biggest challenge for global governance in today's world: how to graduate nations when they emerge from the development process into the industrial world.

References

Antholis, William, and Strobe Talbott. 2007. "Tackling Trade and Climate Change: Leadership on the Home Front of Foreign Policy." In *Opportunity 08*, ed. Michael O'Hanlon. Brookings.

Bull, Hedley. 1977. *The Anarchical Society*. Columbia University Press.

Kindleberger, Charles P. 1986. "International Public Goods without International Government." *American Economic Review* 1, no. 76: 2–13.

Purvis, Nigel. N.d. "Treat Climate Like Trade: The Case for Climate Protection Authority." Unpublished policy brief manuscript.

Ruggie, John. 1983. "International Regimes, Transactions, and Change: Embedded Liberalism in the Postwar Economic Order." In *International Regimes*, ed. Stephen Krasner. Cornell University Press.

Stern, Todd, and William Antholis. 2007a. "A Changing Climate: The Road Ahead for the United States." *Washington Quarterly*, Winter 2007–8, 175–88.

———. 2007b. "Creating an E-8." *American Interest* 2, no.3: 43–48.

Wiener, Jonathan. 2007. "Incentives and Meta-Architecture." In *Architectures for Agreement: Addressing Global Climate Change in a Post-Kyoto World*, ed. Joseph Aldy and Robert Stavins. Cambridge University Press.

C. FORD RUNGE 6

The Climate Commons and a Global Environmental Organization

O ver the period 2000 to 2008, the United States maintained a largely hostile posture toward multilateralism, ranging from military adventurism to
rejection of international norms for human rights and climate change. Its support for the Doha Round of multilateral trade negotiations was undercut by a
failure to live up to global commitments to foreign assistance (the Millennium
Challenge) and protectionist and retrograde 2008 agricultural legislation. If this
experience shows America anything, it is that renouncing its role as a constructive multilateral leader (dating to 1945) has been a disaster for its foreign
policy and the esteem in which it is held.

This desultory period is now at an end, and a fresh start can be imagined in
which U.S. leadership over multilateralism returns. One of the most pressing
challenges relates to global climate change and atmospheric carbon emissions.
This issue is emblematic of environmental challenges that are transboundary
in nature, in which sovereign nations must coordinate effective interventions
with one another. These issues of global commons, of which the climate commons may be the most urgent, may provide an opportunity to push forward an
agenda of institutional change. The specific change considered here is the creation of a Global Environmental Organization (GEO), which can give authority
and responsibility to multilateral climate negotiations.

My argument is that the creation of new multilateral institutions responding to global environmental challenges such as climate change is imperative.
Fortunately, though these institutions would be new, their rationale and even
their structure can be guided by experience with multilateral trade and commercial transactions, notably the World Trade Organization (WTO) and General
Agreement on Tariffs and Trade (GATT) system and the International Labor
Organization (ILO).[1] Just as the GATT/WTO emerged from the postwar con-

1. The present analysis relies on Runge (2001), and reminds one of the Oxford tutor who, com-
menting on his student's essay, noted: "Much of what you say is good and new; unfortunately,
what is good is not new, and what is new is not good." One hopes that while not new, the arguments here are still good.

ferences as a rules-based response to global commercial interdependence, and an earlier ILO from global labor solidarity, so multilateral responses to environmental challenges reflect a growing recognition of nations' ecological interdependence, and a need for rules to coordinate their responses to these global challenges.

This chapter is organized in five sections. First is a brief review of the trade-environment nexus and the rationale for global environmental rules over climate that can coexist with trade regimes. Second is a specific proposal for a GEO, a major responsibility of which would be to coordinate nations' responses to climate change. Third is an appraisal of the implementation of a GEO. Fourth is special consideration of the role of developing countries. A final section offers conclusions.

The Trade-Environment Nexus

In the last two decades, strident criticisms have been leveled at multilateral institutions such as the WTO, describing them as faceless international bureaucracies with programs harmful to the environment.[2] Although hostile to multilateral institutions, these criticisms raise the question: If not *these* institutions, then what others? Though many criticisms of the global economy and global institutions may have merit, it is hard to think of a future in which trade and global institutions, or issues of environment, will play little or no part. Accordingly, the task is to *redefine* objectives in a global economy, and to *restructure* institutions to meet these objectives, especially the urgent challenge of climate change.

The climate change debate has played out, like many other transboundary environmental issues, as a multilateral negotiation over a protocol agreement, the Kyoto Protocol of the Framework Convention on Climate Change (UNFCC) adopted December 11, 1997, which entered into force in February 2005. As of May 2008, 182 parties had ratified the protocol, including 36 developed and 137 developing countries (the United States is the only developed country not to have ratified it). Kyoto commits countries to reduce emissions of carbon dioxide and five other greenhouse gases (GHGs) or to engage in emissions trading to offset such commitments. Partly due to a failure of leadership by the United States, and partly due to exemptions written into the agreement for developing countries, especially large GHG emitters such as China, the protocol has been roundly criticized.

2. For a critique of the 1999 Seattle trade conference debacle in the context of food security, see Runge and Senauer (2000).

This chapter argues that successful multilateral environmental agreements on climate change, as well as many other transboundary environmental issues, may require a framework at a higher level, the result of a meta-agreement on global environmental restructuring. Such restructuring is necessary today at an international level—much as, in the 1780s, the weaknesses of the Articles of Confederation in the United States were increasingly apparent at the national level. Then, Madison, Hamilton, and Jay (writing as "Publius" in *The Federalist*) recognized the need to persuade others that the nation would not endure without substantial institutional innovations. A central element in this school of thought was that free and unfettered commerce should be encouraged between states, coordinated by bodies that derived their authority from the consent of these same states. The concept defended here has similar features, although the states are nations and ecology as well as economy are at stake. The analogy to the framing of the U.S. Constitution is important, because although a GEO may involve ceding some limited national sovereignty over environmental standards, its ultimate purpose is not to impose these standards from the top down. It is to provide greater institutional structure than currently available through treaties or individual multilateral agreements, as well as a nexus of scientific expertise.

One of the reasons for a GEO, although far from the most compelling, is that the GATT/WTO has been unable and largely unwilling to shoulder major environmental responsibilities in conflicts between trade and the environment. This argument has been supported by developments inside the WTO. Its concern over the use of environmental measures as trade barriers has been stung by criticisms of various WTO rulings from environmentalists, notably the "tuna-dolphin" and "shrimp-turtle" cases. Concerned that it show some response to environmental critics, the WTO General Council created a Committee on Trade and the Environment (CTE) in 1995. The CTE was set up to follow the recommendations of the Ministerial Decision on Trade and Environment adopted in 1994 in Marrakech.[3] Though defenders of the CTE claimed that it demonstrated the "greening" of the WTO, it faced a barrage of criticism after the release of its heavily negotiated report to trade ministers in Singapore in December 1996.[4] It had, critics argued, failed to recommend modifications in multilateral trade rules "to enhance a positive interaction between trade and environmental measures." It was precisely the unwillingness of trade ministers

3. See Sampson (2000, 26–29). Shaffer (2001a) emphasizes that the motivation behind the CTE was not simply pressure from groups concerned over the environmental impacts of trade but primarily fears by WTO members, especially developing countries, over the growing number of environmental regulations with potential trade effects. From the point of view of trade ministers the latter dominated the former. (All citations of pages in Shaffer 2001a are to the manuscript version.)

4. Charnovitz (1997).

to redefine trade rules for environmental ends that revealed their essentially (and understandably) conservative posture.[5] Sampson argued that the Singapore report of the CTE showed how wary trade officials were of entering into environmental policy. This suggests that those who have the appropriate environmental expertise—both nationally and internationally—should play a larger role. However, this begs the question of *how* they should play such a role. Since the mid-1990s, a growing number of experts have called for the creation of a GEO.

A Global Environmental Organization

The idea of a GEO is not new. With customary insight, George Kennan argued as long ago as 1970 for the creation of an International Environmental Agency possessing both "prestige and authority" to overcome hundreds of environmental entities, "all of them presenting a pattern too complicated even to be understood or borne in mind."[6] In the 1980s, the late Elliot Richardson argued forcefully in the context of climate change for a permanent environmental multilateral body, whether a "beefed-up UNEP" (United Nations Environment Program) or an entity patterned on the WTO.[7]

In the mid-1990s, both Esty and this author and colleagues proposed a GEO.[8] The main rationale for its creation concerned transboundary environmental challenges, often described as global public goods (or bads), such as climate change, atmospheric ozone pollution, degradation and loss of plant genetic resources, transboundary shipments of hazardous wastes, and threats to endangered animal and plant species. Because they respect no national boundaries, their solution requires joint participation and coordination by sovereign states. In the absence of a "global Leviathan," agreements must be reached that call upon each affected country in the global commons to adopt policies that contribute to a general solution.[9] As noted above, the mode adopted most has been a multilateral environmental agreement (MEA), such as the Kyoto Protocol (1997),

---

5. Sampson (2000, 26–29). Shaffer (2001a, 1).

6. Kennan (1970), quoted by Charnovitz (2002, 18).

7. Richardson's (1992) analysis and call for a Multilateral Environmental Agency was developed in the context of climate change, although the arguments he advanced are general ones.

8. Esty (1994); Runge, Orlalo-Magne, and Van de Kamp (1994)

9. Other early advocates of a GEO were Steven Charnovitz and Jeffrey Dunoff, who suggested modeling it on the International Labor Organization (ILO), as well as Geoffrey Palmer. See Charnovitz (1993); Dunoff (1994); Palmer (1992). See also Esty (1993). For more recent studies, see Whalley and Zissimos (2000) and Biermann (2000). Also see Kaul, Grunberg, and Stern (1999); Sandler (1997); Esty (1994); Runge, Orlalo-Magne, and Van de Kamp (1994).

the Montreal Protocol (1989) on atmospheric ozone, and the Cartagena Protocol (2000) on biosafety and plant genetic modification.

Though it is arguable that MEAs such as Kyoto are an adequate response to these environmental problems, two fundamental questions arise.[10] First, should the MEAs themselves somehow fall outside the trade disciplines of the GATT/WTO system, or (especially when they involve explicit trade measures or sanctions) are they in fact in violation of the principles of free trade? Second, can the hundreds of existing MEAs, and the scores that can be anticipated in the coming decades, including renegotiated agreements on carbon and climate, be adequately managed without creating an institutional umbrella to help oversee the linkages among and between them, and their potential conflicts with WTO rules?

In 2005, Biermann and Bauer collected papers by the growing number of advocates (and some critics) of a GEO in a single volume.[11] First among the advocates was Steven Charnovitz, who has long supported a GEO.[12] It is important to emphasize that despite his intimate familiarity with the legal and economic details of both trade and environmental institutions, his overriding message is the role of a GEO in harnessing the political will to confront transboundary environmental issues, especially if they intersect with trade norms.

As the late Konrad von Moltke noted at the beginning of the decade: "there has never been an occasion when the entire structure [of international environmental management] has been reviewed with a view to developing optimum architecture."[13] A United Nations Task Force observed that environmental activities within the UN (not to mention outside it) frustrate government ministers, especially from developing countries, who do not have the staff to attend the multiplying number of meetings of the various bodies, task forces, and expert groups.[14] As Charnovitz noted, "If an organization chart of world environmental governance existed, its incoherence would be Exhibit A for reformers."[15]

The basic design of a GEO advanced by Runge and colleagues was composed of a Secretariat and a Multilateral Commission on the Environment (MACE) (see figure 6-1).[16] The inspiration for this structure was the North American Commission for Environmental Cooperation, which had emerged from the environmental side agreement to the North American Free Trade Agreement (NAFTA). The GEO Secretariat would be a formal, ministerial-

10. Runge (2009).
11. Biermann and Bauer (2005).
12. For example, see Charnovitz (2002).
13. Von Moltke (2001, 15).
14. UN Task Force on Environment and Human Settlements (1998).
15. Charnovitz (2002, 17).
16. Runge, Orlalo-Magne, and Van de Kamp (1994).

Figure 6-1. The Structure of a Global Environmental Organization

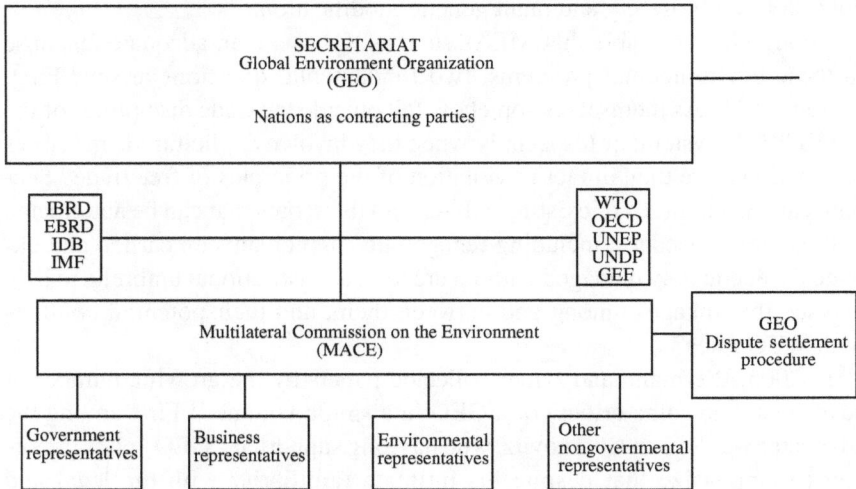

```
                    ┌─────────────────────────────────────────────────┐
                    │                   SECRETARIAT                    │────
                    │           Global Environment Organization        │
                    │                      (GEO)                       │
                    │                                                  │
                    │            Nations as contracting parties        │
                    └─────────────────────────────────────────────────┘

   ┌────────┐                                        ┌────────┐
   │ IBRD   │                                        │ WTO    │
   │ EBRD   │                                        │ OECD   │
   │ IDB    │                                        │ UNEP   │
   │ IMF    │                                        │ UNDP   │
   └────────┘                                        │ GEF    │
                                                     └────────┘
   ┌──────────────────────────────────────────────────────┐   ┌──────────────┐
   │      Multilateral Commission on the Environment       │   │     GEO      │
   │                      (MACE)                           │   │ Dispute settlement │
   │                                                       │   │  procedure   │
   └──────────────────────────────────────────────────────┘   └──────────────┘

 ┌────────────┐  ┌────────────┐  ┌──────────────┐  ┌────────────────┐
 │ Government │  │ Business   │  │ Environmental│  │ Other          │
 │ representatives │ representatives │ representatives │ nongovernmental │
 │            │  │            │  │              │  │ representatives │
 └────────────┘  └────────────┘  └──────────────┘  └────────────────┘
```

Source: Adapted from Runge, Ortalo-Magné, and Van de Kamp (1994).
Note: IBRD = International Bank for Reconstruction and Development—that is, World Bank; EBRD = European for Reconstruction and Development; IDB = Inter-American Development Bank; IMF = International Monetary Fund; WTO = World Trade Organization; OECD = Organization for Economic Cooperation and Development; UNEP = United Nations Environment Program; UNDP = United Nations Development Program; GEF = Global Environment Facility.

level body of government representatives, meeting periodically to affirm certain policies. MACE would be a policy-oriented group of environmental experts drawn from nongovernmental organizations (NGOs), academia, business, and government. Though the representatives to the GEO Secretariat would (like the WTO), be government officials, expert environmental and business involvement was also proposed, similar to the ILO. MACE would thus be composed of a standing group of environmental experts and government and business representatives from all member counties. Its meetings would be open to the public, and it would allow worldwide access to the data and analysis underlying its work. The primary focus of this work would be to propose ways to "harmonize up" national environmental standards, while carefully considering the technical issues and problems of this process for developing countries. MACE would issue regular reports and related documents proposing improved policies, identifying environmental "hot spots," and recommending special projects for national governments. This process would allow for public comments from any group, governmental or nongovernmental. The effect would be to open MACE to full public participation and review.

In the context of climate change and renewed negotiations beyond the Kyoto Protocol, a GEO would backstop and render assistance to the UNFCC, and offer technical assistance especially to developing countries entering into new agreements. It would also help to address potential conflicts between climate conventions and a variety of trade norms, such as the imposition of border tax adjustments to offset the effects of domestic GHG taxes.[17]

The GEO and MACE would work closely with the UNFCC and UNEP to develop funding for environmental projects to upgrade national energy infrastructure, resulting in reduced GHG emissions. National governments would be encouraged to establish an initial tranche of $10 billion for these purposes to operate on a revolving basis. This funding would focus primarily on GHG-reducing projects in those developing countries where national resources for environmental improvements are scarcest. The GEO would not supersede the UN agencies responsible for climate, but it would support them both politically and technically. Eventually, UNEP might be incorporated into the GEO, as Charnovitz suggested, with the GEO styled as a specialized agency under Article 59 of the UN charter, the same authority used to upgrade the UN Industrial Development Organization in 1985.[18]

The GEO would also work jointly with the WTO and the Organization for Economic Cooperation and Development to identify trade measures that threaten environmental quality, and to develop environmental policies that are least burdensome to trade expansion. It would serve as a general "chapeau" for the growing number of multilateral environmental agreements, including the Kyoto Protocol, just as the ILO serves as an umbrella over a large number of special labor agreements and arrangements. Charnovitz proposed as a long-term objective the clustering, harmonization, and codification of many MEAs into a codex of international environmental law, "in which treaties on the same topic are grouped together, duplicative law eliminated, conflicting law reconciled, and eventually the hundreds of MEAs are reduced to a single code."[19] He also proposed that MEAs join this structure voluntarily, so as to avoid conflicts with established governing bodies.[20]

Linking the environmental activities of a GEO to market access and trade reform in the WTO, the Organization for Economic Cooperation and Development, and the multilateral lending agencies would create additional incentives for developing countries to support it. Susskind and Ozawa noted that "environmental negotiations, up to now, have been conducted largely in isolation from negotiations on other international issues such as debt, trade, or security." Linking these issues properly creates a larger negotiating space, and the poten-

17. Frankel (2004).
18. Charnovitz (2002, 20).
19. Charnovitz (2002, 25).
20. This is a problem emphasized by Juma (2000, 13), who argued against a GEO.

tial for mutual gains, because "the goal of a well-structured negotiation is not to encourage compromise but to find ways of ensuring that all parties will be better off if they cooperate."[21] How such linkage occurs is important, and will be considered in the sections below.

Although a highly elaborated plan will require a great deal of analysis and consultation, it is well to ask whether such an organization is really needed, given UNEP and related work by development agencies such as the United Nations Development Program (UNDP) and the Commission on Sustainable Development, as well as hundreds of MEAs. Though supplementing and drawing on the work of these groups, knowledgeable observers and participants still support a GEO.[22]

Precisely because an independent entity such as a GEO is lacking, a greater temptation exists to use trade measures to enforce environmental obligations, harming the world trading system. Trade interests may condemn the use of such measures for environmental goals (such as dolphin-safe tuna), but in the absence of an overarching entity such as a GEO, environmentalists can more easily claim that they have no recourse.

It is naive to imagine that trade and the environment can be entirely disjoint, but the creation of a GEO would assist in separating many issues that do not need to be in conflict. The weaker the perceived ability of environmental groups to influence international policies, the greater their incentive to use "linkage" destructively—to threaten the trading system in order to gain environmental concessions.[23] By drawing environmental expertise and energy into the functioning of a GEO, the GATT/WTO system would largely be left to pursue its own trade agenda, mindful of environmental concerns, and in cooperation with a GEO Secretariat, but not as a functioning "green" trading body.

Together, three arguments thus constitute the core rationale for a GEO: (1) the unwillingness and inappropriateness of the GATT/WTO system as a center for transnational environmental expertise and activity; (2) the widespread number of environmental issues that are inherently multilateral due of their scale and multiple jurisdictions, making them "global public goods" that cannot be adequately managed through existing agencies or ever-proliferating and uncoordinated MEAs; and (3) the logical necessity of separate institutional authority for what are substantively separate environmental problems, which pose a set of targets for policy that require their own instruments at an international level.

21. Susskind and Ozawa (1992, 153). See also Barrett (1992).
22. For a discussion of the role of the UN agencies in global environmental affairs, see Thacher (1992). A cautionary note on the need for new institutions in the context of NAFTA is given by Mumme (1992). The range of supporters for a GEO has nonetheless continued to expand.
23. Hauer and Runge (1999).

Implementation and Policy Constraints

There are important reasons why a GEO may be beyond the reach of the world's governments and their leaders. The first is that it will be opposed as unnecessary—that existing institutions, suitably augmented, are adequate to respond to transnational environmental challenges. The second is that it is unwieldy—another international bureaucracy that may prove just as unresponsive as existing ones to the concerns and interests of member states and may actually challenge their sovereignty over national environmental issues. The third, and most potent, is that its creation would reflect the same "rich man's club" priorities—which, in the view of many developing countries, have dominated the GATT/WTO system, tilting its functioning toward priorities of the North rather than the South.

The first argument is that the panoply of existing UN agencies, NGOs, and MEAs together constitute a sufficient response to transboundary environmental issues. These include UNEP, the UN Commission on Sustainable Development, and the hundreds of MEAs noted above. Others include the UNDP, the World Bank, the World Meteorological Organization, and the Global Environmental Facility. In addition, a growing number of NGOs—such as the World Resources Institute in Washington, the World Wildlife Fund, and the Center for International Environmental Law—have become active participants in the trade/environment agenda. Juma notes that because of the diversity of environmental problems, specialized institutional responses are often required, reflected in the MEAs and other agreements that deal with these questions issue by issue. Though coordination may be desirable, in his view, "centralization" is not.[24]

The second claim leveled at a GEO is that it is likely to be an unwieldy and unresponsive international bureaucracy of its own, which simply adds another layer to the many and diverse responses to global environmental problems noted above. Below the surface of this argument are GEO opponents that are relatively comfortable with their influence over existing institutions, and that fear that they would lose this influence in a new body. These groups include not just bureaucrats at bodies such as UNEP but also state agencies and NGOs. It is arguable that member states of any multilateral body, along with stakeholders such as environmental NGOs, seek to capture it for their own purposes. Such investments in capture, once made, are defended against new and uncertain prospects.[25]

24. Juma (2000) asserts that Esty's arguments in favor revolve around "administrative efficiency" claims. A careful reading of Esty (1994), and the arguments developed here suggests that administrative efficiency, even if improved by a GEO, is not a central argument in its favor, especially in light of the struggles it would face from existing UN agencies.

25. See Shaffer (2001a, n. 27). Another analysis of the links from domestic political interest to trade and environment issues at the international level is given by Shaffer (2001b).

A GEO that is less subject to capture, and therefore "unresponsive," is also less subject to special interests. By increasing the scope for coordinated approaches to global environmental issues, a GEO may reduce opportunities for exercising such influence, and thus arouse concerted opposition from defenders of the status quo.

A related issue concerns the many national agencies and ministries to which existing MEAs and agreements are tied back. At an administrative level, the authority for various aspects of international environmental policymaking emanates from these different parts of national governments. In the United States, while the Executive Office of the President is ultimately responsible, duties for international environmental policy are parceled out across a large number of executive agencies, from the Environmental Protection Agency, to the National Oceanic and Atmospheric Administration (part of the Department of Commerce), the State Department, the Department of Energy, the Office of the U.S. Trade Representative, and the Department of Agriculture, among others. Each agency will defend its role in status quo agreements against any "coordination" that diminishes it.

The third and most potent forces arrayed against a GEO are developing countries that are convinced that it may force Northern priorities on Southern interests. These include not only environmental goals regarded as lower priorities in developing countries but also trade protection in "green" disguise. As Juma notes, "Many developing countries are concerned that a new environmental agency would only become another source of conditions and sanctions."[26] These concerns were amply revealed in the WTO's CTE. In opposing even the formation of the CTE, spokespeople for the Association of Southeast Asian Nations, such as Thailand, and other less-developed-country representatives from Morocco, Tanzania, and Egypt, all questioned the need for it.[27] Shaffer noted that none of them wanted "to be pressured into signing an environmental side agreement analogous to NAFTA's." When the CTE agenda was finally settled, it reflected a cluster of issues that linked less-developed-country environmental initiatives to the achievement of expanded access to Northern markets.[28]

However, even these concessions did little to assuage nervousness by developing countries concerning the possible growth of environmental conditionality. Of particular concern was the widespread sense that environmental demands would join similar demands by labor interests in the North to justify shutting off developing countries' market access, a view reinforced by the political

26. Juma (2000, 15).
27. Shaffer (2001a, 10).
28. Shaffer (2001a, 14).

alliances struck between greens and labor in opposition to trade liberalization. Discussing the idea of opening the Article XX exceptions to broaden allowances for environmental measures, for example, Brazil's deputy permanent representative to the WTO stated in 1998 that "we [developing countries] cannot be in favor of a change in Article XX. We think that this would create an imbalance in terms of a whole set of disciplines and commitments and would set a precedent for other issues." As Shaffer notes, the other issues he had in mind were trade restrictions based on "unfair" labor standards.[29] It is particularly noteworthy that Mexico, after acceding to the NAFTA environmental side agreement with the United States and Canada, led the opposition to many U.S. proposals in the CTE. When the U.S. delegation questioned whether Mexico's representatives to the CTE were speaking for the Mexican government, Mexico City quickly confirmed that these opposing views were indeed official positions.[30]

Together, these three claims pose serious challenges to successfully launching a GEO. To recapitulate, they are, first, that existing bodies and agreements respecting international environmental issues are adequate, and do not require a centralized overarching entity. Second is that a GEO would be unwieldy, simply adding another layer of bureaucracy to existing agencies and groups, most of which will oppose any attempts at coordination that diminishes their influence. Third is that most developing countries will oppose any new body that could pressure them to conform to higher environmental norms or standards or risk reduced access to Northern markets. Hence, any successful argument in favor of a GEO must demonstrate that existing arrangements are *not* in fact adequate and that coordination may not imply centralization; that a GEO can be implemented in a way that accommodates and complements existing institutional arrangements; and that less-developed countries' suspicions and reservations can be overcome.

It is clear that the creation of a GEO would pose difficult issues of implementation. Among them are: (1) What duties of existing bodies would be assumed by a GEO, and what would these bodies then do? For example, the GEO should add value to the IPCC by offering scientific and policy advice and direction on climate change, without trampling on the IPCC's role. (2) What new responsibilities would be assigned to a GEO by its members, and by whom would these duties be performed? If the GEO is as lightly staffed as the WTO Secretariat, for example, how can it manage to draw on university research on climate change that can be brought to bear on negotiations? (3) What would

29. Shaffer (2001a, 21).
30. Interview with Ricardo Barba, deputy permanent representative to the WTO from Mexico, quoted in Shaffer (2001a, 26).

be the relationship between a GEO and the WTO? Would consultations between the two be informal or formalized? Though no definitive answers can be given to these complex legal and administrative questions in a brief treatment, some general comments are in order.

First, it is probable that a GEO would eventually assume some of the responsibilities of UNEP and the Commission on Sustainable Development (CSD). This is in part, Esty argues, because UNEP as an agency tries to do too much. The CSD is similarly overstretched. In addition, there are responsibilities of the UNDP and the World Bank related to environment and development in which the GEO might assist, assuming that development projects remained the province of these groups. The GEO could, for example, assist in the planning of expanded irrigation schemes involving interbasin and/or international transfers of water so as to minimize environmental disruptions. A major function of the GEO would be to provide a transparent source of information on global environmental issues, assisting what is now often the task of NGOs. Esty notes that someone attempting to track environmental decisions at the WTO "would find out a great deal more by reading newsletters from the World Wildlife Fund then communiqués from the Office of the U.S. Trade Representative."[31] Although groups such as UNEP and even some NGOs might feel threatened by a GEO, it is probable that enough work will remain to keep every group fully engaged in international environmental affairs. However, to the extent that budgetary resources are drawn off existing agencies and programs to support a GEO, internecine competition will be intense. This is why the proposed $10 billion infusion is necessary, which should be linked to resources drawn from the World Bank's Climate Investment Fund.

As noted above, a GEO would offer a "chapeau" for the growing number of MEAs, especially in the context of dispute settlement, and in coordinating the MEAs. An analogy is the role played (since 1967) by the World Intellectual Property Organization (WIPO), headquartered in Geneva. The WIPO was established in part to help unsnarl the "treaty congestion" that surrounded intellectual property and patent rights, and to help rationalize and coordinate these efforts.[32]

It is also arguable that a GEO would help to offset the perception in developing countries that MEAs and exceptions granted to WTO contracting parties under GATT Article XX or other headings are heavily tilted in the direction of the Northern states. The Indian NGO Centre for Science and the Environment,

31. Esty (1999, 1564). Indeed, the total budget resources devoted to these efforts by NGOs considerably exceed those of subagencies of UNEP responsible for environmental information. Shaffer (2001a, 32 n. 97) notes that Greenpeace's annual income in 1998 was $125 million, and that of the World Wildlife Fund was $53 million.

32. See Esty (1994, 96).

for example, "characterized the use of trade measures in MEAs as an inequitable lever available only to stronger countries."[33] As noted above, so long as this perception continues, Southern countries will remain skeptical of global environmental initiatives. Yet a GEO may be precisely the mechanism needed to give added weight to these Southern concerns.

One of the most pressing and unmet needs to which a GEO could contribute is preparation and technical support available to developing countries in the formulation of development and environmental initiatives. If a GEO is to succeed, it must treat these needs as of paramount importance. In particular, a GEO should take as its responsibility the implementation of the primary principles emerging from the 1992 Rio de Janeiro Declaration on Environment and Development (ostensibly the current responsibility of the CSD):

—that developing and developed countries have differing responsibilities to enact domestic measures to protect the environment;

—that international transfers are necessary to assist developing countries to upgrade their environmental protection measures;

—that unilateral measures are to be avoided.[34]

In the context of a GEO, these three principles imply that a form of "special and differential treatment" in environmental policies should be part of an international body of multilateral environmental rules, in which the differing capacities of the North and South to mount programs of environmental protection are realistically acknowledged. The resources to undertake environmental programs are substantial, and it is unreasonable to expect countries at early stages of development to give them priority. Just as unilateralism in trade policy is ultimately self-defeating, so it is where transborder environmental issues are concerned. A GEO would not require all national environmental measures to be subjected to oversight, but where these measures affect the "global commons," multilateralism should provide a foundational principle. Finally, the GEO/WTO interface will be all-important. Perhaps, paradoxically, if it is to take environmental pressures off the WTO, a GEO should be located in Geneva. There, it could assist the WTO (analogous, again, to WIPO), and it would be situated to work in cooperation with the World Health Organization and the growing number of environmental NGOs that have found it useful to use Geneva as a base.

A last set of implementation issues concerns timing and phasing. It is very unlikely, given the problems and potential opposition facing a GEO, that it could

33. Quoted by Shaffer (2001a, 36). Biermann (2000, 25–26) argues that improving technology transfers to developing countries for environmental improvements could be a major GEO function.
34. Quoted by Shaffer (2001a, 47 n. 132).

be implemented in a single "grand stroke," nor should climate change negoti-ations be delayed while waiting for the GEO to be set up, which could occur as a parallel and reinforcing process. Achieving new climate rules, however, will be more likely to emerge from a more general agreement to reform and revitalize all the major environmental institutions in a single exercise analo-gous to the momentous postwar conferences of the 1940s. Another possibility would be to launch a GEO as part of a refreshed multilateral trade negotiation (MTN) after the WTO Doha Round. Such an outcome assumes that such a round can be successfully negotiated by a new U.S. administration, with side nego-tiations on a GEO contributing part of the final package. A third approach would be to open negations on a GEO as a multilateral environmental effort, *linked to, but separate from*, MTNs. This would follow, in general form, the NAFTA side agreement model and would place GEO talks on a separate path. A dif-ference might be that whereas the NAFTA environmental side agreement would surely not have succeeded had NAFTA failed, GEO talks might proceed and even succeed without a successful MTN. However, for a variety of reasons, related especially to the market access requirements of developing countries, this outcome seems unlikely.[35]

Sampson advocated the use of an "eminent persons group," for global envi-ronment issues on the model of the Leutwiler Group in the run-up to the Uruguay Round.[36] Such a group might demarcate subnegotiating groups sim-ilar to the Uruguay Round to take up issues such as MEAs, climate change specifically, and exceptions to trade agreements for environmental measures under Article XX.[37]

35. Shaffer (2001a, nn. 110, 111) discusses the five NGO Symposia held by the WTO. See Ricupero (1998).

36. Sampson (2000).

37. Biermann (2000) offers three "models" for a GEO: a cooperation model, a centralization model, and a hierarchization model. The cooperation model would essentially retain all existing bodies in their current state but would elevate UNEP to a leading and coordinating role, becom-ing in effect a GEO. The centralization model would grant greater authority to a central institutional actor (again, probably UNEP) to oversee and direct the environmental activities of other UN bod-ies. This model would make the GEO similar to the WTO, and would bring MEAs into a reporting relationship with the GEO. This GEO would have a double-weighted voting procedure, in which decisions would require a two-thirds majority of both developed and developing countries. The hierarchization model would grant the GEO enforcement authority like the Security Council, and would constitute the most dirigiste of the alternatives—approaching a global government. Of the three alternatives, a hybrid of the first two, "cooperation" and "centralization," is closest in spirit to that discussed here. One might call this hybrid the "coordination model."

The Role of Developing Countries

Considerable attention has already been given to the role and interests of developing countries in successfully negotiating a GEO. As a group, the developing countries felt left out of the many rewards promised in return for their support of a Uruguay Round Agreement and sought relief in the now nearly moribund Doha Round. They are equally skeptical of making new commitments to global carbon and climate agreements. Since the completion of the Uruguay Round, global trade has risen significantly faster than gross domestic product (GDP), but the share of developing country exports has fallen relative to that of the United States and European Union. However, if climate commitments are tied to expanded market access to the North, the developing countries may show renewed interest in both trade liberalization and environmental protection. Brazil made market access in agriculture a precondition for the success of the Doha Round. India argued that the commitments it made on intellectual property rights in the Uruguay Round have not been matched by the expanded market access in agriculture and textiles that it expected to ensue.[38] Market access for developing countries probably offers the only prize sufficient to induce these countries to take a GEO and climate and carbon commitments seriously, and would undercut the suspicion that a GEO is a protectionist Trojan Horse, while underscoring economic growth as providing the wherewithal to make the investments required to protect their environments.

A last, and especially thorny, issue concerns the potential role of a GEO as an imposer of sanctions and conditionality on countries unwilling or unable to comply with norms or standards, such as global climate-induced reductions in air pollution. Sanctions are an old and contentious issue in trade policy.[39] Most multilateral agreements, including a renegotiated climate agreement, must carry penalties for noncompliance. However, there is no reason why such penalties need to take the form of *trade* sanctions, as opposed to fines, denial of voting rights, or other measures decoupled from trade itself. This argument can be employed in order to *separate* environment from trade measures, reducing the potential use (and abuse) of trade sanctions to enforce multilateral environmental compliance.

The experience of sanctions in trade policy suggests that they are far less important to the maintenance of world trade rules than the dispute settlement

38. See Finger and Schaler (2000); Ricupero (2000).
39. See Hudec (1993). In the case of environmental obligations and sanctions, see Barrett (2003). In a theoretical article, Barrett (1997) argues that the existence of trade sanctions in environmental agreements may nudge countries toward compliance, so long as a sufficient number of countries are committed to their enforcement.

mechanisms of the WTO, which are in turn modeled on those established in 1919 by the ILO.[40] If a GEO were to come into being, the opportunity to create a separate environmental dispute resolution process relating especially to carbon and climate might be of special importance. Such a process would also allow NGOs to enter disputes as "friends of the court." Though not a formal sanction, NGOs' capacity to focus international attention on countries found to have violated global norms on climate and carbon could have important effects on compliance.

Conclusions

Despite the necessary limits of this analysis, a number of conclusions emerge. The first is that a GEO could help to frame and promote new negotiations on carbon and climate, promoting changes in both the trading system and the global environment. To the trading system, it offers the opportunity to disentangle trade from environmental matters, allowing the WTO to focus where it should: on the expansion of market access and reductions in trade protectionism, saving attention for climate measures only in cases of obvious trade distortion. A GEO could be of considerable assistance to the WTO in clarifying where environmental exceptions to the GATT articles were justified (under Article XX or other headings) in cases affecting carbon and climate, and providing guidelines for minimally trade-distorting MEAs beyond a renewed Kyoto Protocol. At the same time, a GEO could help fill the institutional gap in dispute resolution and coordination surrounding the many MEAs and institutions now responsible for global environmental issues, such as the UNFCC. This coordination need not imply centralization, nor the usurpation of authority from these bodies or national governments.

Second, a GEO could channel needed attention to a wide range of global public goods and global commons issues—notably climate change, but also ozone depletion, biodiversity and overfishing. These issues are in need of greater focus and attention independent of the trading system, suggesting a need for separate multilateral instruments such as a GEO.

Third, overcoming opposition to a GEO will require a twofold undertaking that involves the politics and postures of both developed and developing countries. In the North, opposition to multilateral institutions generally—arising from both the right and the left—must change. Conservatives will need to overcome

40. Steven Charnovitz, personal communication, January 10, 2001. See Charnovitz (2000). For reviews of different types of carrots and sticks, see Esty (1994, appendix) and Chang (1997). For a broader assessment of compliance issues, see Chayes and Chayes (1995).

their distrust of global environmental initiatives. The environmental left, mean-while, must overcome its strident opposition to all things multinational. In developing countries, climate change commitments can now be deflected by claims that such issues are rich countries' problems. Unless a GEO clearly offers specific commitments to special and differential treatment for developing coun-tries, expanded technical assistance, and ample representation, it will be easily discredited as a form of environmental conditionality and a disguised mecha-nism of Northern influence.

Fourth, it is unlikely that developing countries will find a GEO attractive unless it is linked to commitments for expanded market access, especially in key areas such as agriculture and textiles. This suggests a model in which a GEO is *linked* to but *separate* from trade negotiations. The virtue of linkage is that developing countries will see that market access will enable them seriously to contemplate environmental improvements in the context of economic growth. The virtue of separation is that a successful MTN will not have to internalize questions of multilateral environmental policy.

The overall conclusion is that despite serious hurdles, a GEO can be envi-sioned that is both protrade and proenvironment, strengthening commitments to carbon reduction and climate, while carving out new areas of international environmental competency. Achieving this vision will be difficult, but it is this author's view that we are condemned to succeed.

Comments

Comment by Colin I. Bradford Jr.

Both chapters 5 and 6 are of great interest and relevance to the issues in this volume, and both make excellent arguments and raise issues in a stimulating fashion. In the name of effectiveness, both chapters raise concerns about the dangers of having too many countries involved in specific issues and raising too many issues in a single forum.

Energy as the Door into Climate Change

With due deference to all those—including Ford Runge, in chapter 6—who make a good case for a Global Environmental Organization (GEO), with whom I am sympathetic, my view would be that for the purposes of negotiating a climate change framework for the post–Kyoto Protocol period, it would be better to begin with a focus on the energy-climate change link and to approach the issues of global climate change through energy rather than through the environment, given all its multiple meanings and dimensions. Approaching global climate change through energy would have a greater chance of engaging leaders from energy companies, banks, investment firms, technology companies, and other private sector leaders, who ultimately bear the responsibility for meeting the energy supply challenges facing the globe over the next fifty years, as world population will increase by nearly 3 billion people.

I think a case can be made that "the environment" needs a GEO, just as trade has the World Trade Organization (WTO), health has the World Health Organization, finance has the International Monetary Fund, and development has the World Bank. "The environment" is immensely broad in its scope and involves an enormous number of issues, which would benefit from an integrated approach and a single focal point to enable the world community to work effectively on these issues. But negotiating an energy–climate change framework requires a more flexible and agile mechanism, such as that developed in chapter 5 by William Antholis, more than it needs a new, permanent umbrella agency to move on this specific issue over the next few years. A new formal organization is more the work of a generation, as Antholis points out, than it is a solution to our current challenges. I will return to this point at the end of this comment.

Governance Parallels between Trade and Climate Change

Chapter 5 gives an excellent synoptic summary of the parallels between the evolution of the international institutional arrangements for trade and those for global climate change. The story is insightful and useful for providing a pathway for evolutionary institutional change. The similarity between the universal membership forums for the discussion of trade and climate change (the United Nations Conference on Trade and Development, and the United Nations Framework Convention on Climate Change), the intermediate steering groups (the Quad, or Group of Eight, G-8; and the Major Emitters' Group, E-8, or Group of Thirteen), and the global governance institutions (the WTO, and, eventually, a GEO) sets a relevant and practical backdrop for the current governance issues related to climate change. This backdrop provides the sense that progress is indeed possible based not only on hope but also on experience.

Governance Parallels between Summit Reform and Climate Change

With our Canadian colleagues at the Centre for International Governance Innovation in Waterloo and the Centre for Global Studies in Victoria, we at Brookings have been working for the past four years to try to advance summit reform. The crux of this work fits the pattern laid out in chapter 5. The central issue is how to find an intermediate global steering group that is between the universal membership of the United Nations and a group, in this case the G-8, that is too small to be representative of the global community as currently constituted and emerging in the future. This collective effort has concluded, after surveying a number of global challenges, that energy security and climate change together have the greatest leverage to compel summit reform and broaden the group of countries included in the annual meetings of heads of state that seek to address global issues. As a result, we ended up in the same place as Antholis, but by a different route. We have concluded that a summit grouping composed of at least the G-8 plus five emerging market economies

(China, India, Brazil, South Africa, and Mexico), and possibly a slightly larger grouping, would provide significantly greater representative legitimacy than the G-8 alone and still be small enough to be able to reach closure, make effective decisions, and thoroughly discuss the issues. Thus, it would be both effective and legitimate, not one without the other.

Small Size an Issue for Efficiency, but Not for Representativeness Legitimacy

Therefore, though there is convergence between the summit reform efforts sketched above and the governance group proposed by Antholis, there are some caveats. I think it is fair to say that Antholis is more concerned about efficiency than legitimacy, and about effectiveness than representativeness, though he clearly is concerned about both. His concern is understandable. It is also a strong undercurrent in the summit reform debate. Many observers are concerned that a leaders'-level Group of Twenty (L-20), parallel to the finance ministers' G-20, would be too large, even though most finance ministry officials with whom we have talked find that the G-20 finance ministers' group works extremely well. In part, because of these concerns, the consensus in our discussions over the last several years has dropped back from advocacy of an L-20, originally proposed by the former Canadian prime minister, Paul Martin, to something larger than the G-13 implied by the current practice of partial summit meetings of the G-8 plus 5.

It is important to be aware that the E-8 described by Antholis is in reality a group larger than eight. The E-8 is composed of the United States, the European Union, Japan, Russia, China, India, Brazil, and South Africa. Realistically, in the near term, the Europeans, very regrettably, are not yet ready to have themselves represented at summits or in any of the major international institutions by a single EU representative. So Antholis's E-8 becomes an E-11 by seating France, Germany, Italy, and the United Kingdom, instead of just the EU. The E-8 (= E-11) also leaves out Canada and Mexico from the G-8-plus-5 group; adding them back, as it were, into a mix that many would consider the current consensus, brings the E-8 to an E-13. Then there is the issue that Australia, Indonesia, and South Korea, in addition to the thirteen, are also among the sixteen in the major emitters' group, which would bring the energy–climate change negotiating group to sixteen, not eight. Give or take a few, this is more likely to be the size of the group that can be both more effective and more legitimate, on this issue and perhaps on others, than a smaller group. One way to say it is that an E-16 would be better than an E-8, not just bigger.

This result is in accord with a recent survey I undertook for a chapter on the United States and summit reform for a forthcoming book from the Centre for International Governance Innovation.[1] Seventy-six people responded to a questionnaire with more than fifteen questions in early 2008. Thirty of the respondents were officials from sixteen G-20 countries, either in Washington or other capitals, and forty-six were experts (academics, researchers, think tank scholars) from the sixteen countries. Of the seventy-six, twenty-six were from the United States (mostly experts), and fifty were from the other fifteen countries (most of the G-8 plus 5, along with Australia, Argentina, South Korea, and Turkey). On the issue of summit size, between 90 and 100 percent of both those from the United States and from other major countries viewed "the recent evolution of the G-8 into a meeting of leaders from G-8 industrial countries plus 5 'outreach' countries" as *important, positive*, and *necessary*, but fewer than 25 percent of both groups found it to be *adequate*. When asked "Would you favor other changes in the leaders-level summit, beyond thirteen?" 79 percent of the U.S. respondents and 85 percent of the respondents from the fifteen other major countries concurred.[2]

Between Intermediate Group and Formal Organization

Another cautionary note I would strike is to raise the question of whether new international institutional arrangements would be needed to guide the major countries through the energy–climate change issues over time. The global energy summit in Jidda in June 2008 revealed the fissure in the international community between energy-producing and energy-consuming nations. It also reinforced the fact that the oil producers meet under the aegis of the Organization of the Petroleum Exporting Countries and the oil-consuming countries meet under the aegis of the International Energy Agency at the Organization for Economic Cooperation and Development (OECD). Yet there is no global focal point for oil or for energy in general, which includes all the major players in the global energy market. Most of the big private and public sector actors seem to prefer this fractured, fragmented, uncoordinated system to anything more organized because they are big and thus have first-mover advantages, profit from information gaps, and want to strike deals without having to put all their cards on the table.

There is a long distance between a summit grouping and a formal "treaty organization," as Antholis calls WTO/GEO-type institutions. The question is

1. Bradford (2008).
2. See Bradford (2008).

whether the issues of energy and climate change would not benefit over the long run from a small secretariat that would present questions for private and public leaders to discuss, especially those related to long-term energy supply, rather than leaving these lumpy, risky investments with long gestation periods to the haphazard opportunities that present themselves to different actors at different moments in different geographical areas. There is a global energy market today, but there is no focal point for energy in the global landscape. And though there is no single silver bullet of governance innovation for global energy, one idea would be to build on the OECD's assets to create a Global Energy Council, which is elaborated elsewhere.[3]

Comment by Daniel W. Drezner

As someone who knows more about international trade policy than the environment, it is difficult to read chapter 5 by William Antholis and chapter 6 by C. Ford Runge without a rueful smile. No doubt, the global institutional machinery for dealing with trade seems to rest on a stronger foundation than that for dealing with global warming. Nevertheless, as I write this in the middle of 2008, the political trend lines on the two issues are moving in opposite directions, both globally and in the United States. The public consensus on the causes and outcomes of global warming has never been greater.[1] Both major party nominees for the presidency in the United States pledged to create a cap-and-trade system for reducing emissions of carbon dioxide and other greenhouse gases. The congressional wrangling over the Warner-Lieberman Bill, though frustrating, revealed a majority in favor of *some* action on the issue. In contrast, the bipartisan consensus on trade liberalization is now officially dead.[2] In the spring of 2008, the Democratic Party candidates for president squabbled over who disliked the North American Free Trade Agreement more. The Doha Round of multilateral trade negotiations has stalled out. The bilateral free trade agreements with South Korea, Colombia, and Panama are unlikely to be ratified. Environmentalists should be wary of copying the trade regime, if these are the political results!

I applaud both Runge and Antholis for crafting arguments that are fully cognizant of the domestic political realities with respect to environmental protection. Antholis's suggestion to ratify environmental accords as congressional-executive agreements rather than as treaties is a productive step forward. I certainly admire

3. See Bradford (2007).
1. Evans and Steven (2007).
2. Drezner (2006).

both authors' optimism in the face of such political realities. I am less convinced, however, that these two chapters grasp *all* the sources of international cooperation that led to the current trade regime. Both chapters postulate that the origins story of the General Agreement on Tariffs and Trade / World Trade Organization (GATT/WTO) system can be replicated in the environmental realm. This kind of analogical reasoning can be prone to error, however. Runge and Antholis fail to recognize how 1940s trade politics differs from twenty-first-century environmental politics. The distribution of power, the calculation of interest, and the institutional framework are all different—and on each dimension, the current situation reduces the likelihood of creating a Global Environmental Organization (GEO) or a General Agreement to Reduce Emissions (GARE).

Both Antholis and Runge discuss the creation of the GATT regime from an international legal perspective—that is, it was a negotiation among sovereign states that existed as equals on the global stage. Their discussions are not inaccurate, but they do omit the crucial role that the distribution of power played in the creation of the GATT. The United States was the unparalleled hegemon in the late 1940s, responsible for close to half of the world's output. After World War II, the United States was bound and determined to foster an international economic order that would prevent the closure of the 1930s. This included a trading system to ensure that all participating members received nondiscriminatory treatment in traded goods.

This combination of power and interest on the part of the hegemon guaranteed that the GATT would be created. To reassure nervous allies, the United States was willing to "bind" itself to the GATT, demonstrating that the rules of the game would be respected.[3] The United States also grandfathered in the British imperial system, ensuring that the most important "follower" state was willing to cooperate. In both cases, American primacy was so great that Washington was willing to incur short-term costs for long-term gain. As hegemonic stability theory suggests, the United States also provided the lion's share of global public goods—the reserve currency, a security umbrella, foreign aid transfers—that kept the system functioning.[4]

The current situation looks rather different. The United States remains the most powerful country in the world, but outside the military sphere it is merely the first among equals. China has supplanted America as the biggest emitter of greenhouse gases. From an economic perspective, we are witnessing a transition from a bipolar world (the United States plus the European Union) to a multipolar world (the members of the Organization for Economic Cooperation and Development plus Brazil, Russia, India, and China, known as the BRICs).

3. Ikenberry (2000).
4. Frieden (2006).

International relations theory is not sanguine about what this means for international cooperation. In theory, a concert of great powers can still foster cooperation.[5] Possible does not mean likely, however.[6] In practice, as the number of powerful actors increases, the likelihood of meaningful cooperation declines.[7] The unending Doha Round of multilateral trade negotiations is obvious evidence of this. The current distribution of power does not make the creation of a GEO or GARE impossible—but it does make it much more difficult.

The constellation of state interests on global warming is also different from trade. The GATT framework thrived because this noncommunist institution could simply bypass the Soviet bloc. At the outset, only like-minded states participated in the GATT system. Over time, what started out as an exclusive club transformed itself into a universal-membership institution.

This growth pattern worked for two reasons. First, nondiscriminatory trade rules are not a perfect public good; they are at least partially excludable. If states did not wish to join the GATT system, then the GATT's members were under no obligation to bestow most-favored-nation trading status upon them. Access to the dispute settlement mechanism was also blocked. The GATT's member states had the option of erecting higher trade barriers against nonmembers. Because trade liberalization contributed to greater economic growth (through trade creation) as well as the relative reduction of trade between members and nonmembers (through trade diversion), the opportunity costs of noncooperation were manifestly clear. In this manner, the members of the Quad (the United States, Japan, Canada, and Europe) were able to create a "club standard" on trade—in which other countries had little choice but to join.[8]

Although trade might generate partially excludable benefits, the mitigation of climate change is essentially a pure public good. With this kind of public good, each country is best off when it can free ride off every other country's costly steps to reduce carbon emissions.[9] Global warming represents a classic tragedy of the commons / *n*-person prisoner's dilemma problem in international relations. To be sure, mutual cooperation is better than mutual defection, but unilateral defection is the best of both worlds. This incentive structure requires the creation of a different mix of incentives and punishments to ensure effective cooperation.

International relations theory suggests two mechanisms to ensure cooperation: the logic of appropriateness, and the logic of consequences.[10] The former

5. Snidal (1985).
6. Axelrod and Keohane (1985).
7. Krasner (1976).
8. Drezner (2007a); see also Gruber (2000).
9. Barrett (2007).
10. March and Olsen (1998).

relies on legal or normative obligation to deter noncooperation. Defecting states are viewed as violating a commonly accepted norm. Certainly, the normative pressure to "do something" about global warming has been on the increase. The problem is that developing countries—which must participate for mitigation to be appreciable—can deploy a counternorm of fairness. They can argue that because the economies belonging to the Organization for Economic Cooperation and Development were historically responsible for the bulk of carbon dioxide emissions, they should be responsible for shouldering the burden of ameliorating the problem. With competing norms at play, it is unlikely that a logic of appropriateness will work on its own.

The logic of consequences involves the creation of material rewards and punishments to encourage more compliance. As the other chapters in this volume suggest, however, the utility of trade sanctions or border taxes in punishing noncompliance is problematic from either a legal or a welfare perspective. This leaves the creation of excludable incentives for nations that choose to join a GARE, a GEO, or an emitters' grouping.

This leads to a mildly perverse suggestion. Because the effects of mitigation cannot be excluded, any new club should create an excludable benefit that reduces the adaptation costs with respect to global warming. The expert consensus on global warming is that regardless of what is done to mitigate its effects, adaptation to elevated levels of greenhouse gases will be required. Furthermore, this burden will fall disproportionately on the developing world. If a GARE or a GEO proffers an adaptation fund of some sort, it *could* encourage the necessary levels of cooperation.

Perhaps the biggest difference between the GATT's origins and the attempt to develop a parallel environmental organization is the existing institutional environment. The GATT was created in an institutional vacuum—there were no other significant trade institutions in existence in 1947. As John Ikenberry has observed, the principal efforts to craft "constitutional orders" have followed major power wars.[11] These wars discredited the legitimacy of the old order and generated decisive shifts in the global distribution of power. This allowed the military victors to write the rules, once the slate was wiped clean of preexisting institutions. At present, however, the slate is not even close to clean. There has clearly been a steady increase in the number of conventional intergovernmental organizations, autonomous conferences, and multilateral treaties—even in the environmental realm.[12]

What will be the effect of adding a GEO or a GARE to this unclean slate? Liberal internationalists believe this trend to be a good thing. The editors of

11. Ikenberry (2000).
12. Drezner (2008).

Legalization and World Politics observe approvingly that "in general, greater institutionalization implies that institutional rules govern more of the behavior of important actors—more in the sense that behavior previously outside the scope of particular rules is now within that scope or that behavior that was previously regulated is now more deeply regulated."[13] In their chapters, both Antholis and Runge express the belief that the new organizations they conjure into existence will act to coordinate the existing morass of environmental institutions.

As international regimes proliferate, overlap, and impinge upon each other, there are several reasons to believe that the whole might be less than the sum of the parts. Institutional proliferation allows states to engage in strategic forum shopping. This can be seen in the trade realm, where the explosion of free trade agreements has weakened the GATT/WTO regime.[14] The existence of nested and overlapping governance arrangements makes it more difficult to detect opportunistic defections from existing regimes. The creation of legal mandates that conflict over time can weaken all actors' sense of legal obligation.

Given the sharp distributional implications of global warming, the creation of new environmental organizations—particularly a GEO—risks the creation of "sham governance." Governments will agree to a notional set of global regulations with weak or nonexistent monitoring or enforcement schemes. Sham governance is politically useful to states of all stripes, because it permits governments to claim the de jure existence of cooperation, even in the absence of effective de facto enforcement. Thus, a GEO could act to relieve or redirect any domestic or civil society pressure for significant global regulations. And it could also create path dependencies in governance institutions that would cast a shadow over future governance efforts.[15]

An alternative possibility is to reform existing institutions to incorporate global warming as an issue area. Ideally, the way to do this, as Antholis hints in chapter 5, is to expand the Group of Eight (G-8) to include the BRICs—Brazil, Russia, India, and China—and also South Africa. Unfortunately, these countries are wary of joining a club that they view as dominated by the advanced industrial states. Furthermore, international relations theory does not provide much guidance about how to reform existing global governance structures. Reform efforts to date are hardly encouraging on this point.[16]

This logic leads to a concluding and cautiously optimistic note. If policymakers can clear the hurdle of expanding and adapting the G-8 to address global

13. Goldstein and others (2001, 3); also see Slaughter (2004).
14. Bhagwati (2008).
15. North (1990); Raustiala and Victor (2004).
16. Drezner (2007b); Bradford and Linn (2007).

warming as an issue area, the chances for success would be quite decent. Because the G-8 tackles a diverse set of issues, the possibility for tactical issue linkages and cross-issue bargaining increases.[17] Global warming, by itself, is a Gordian knot of distributional issues. But if it can be connected to other matters of high and low politics, a grand bargain would be possible.

17. Adding global warming to the G-8 agenda would also address the bureaucratic politics problem identified by both Runge and Antholis. Requiring policy principals to address global warming policy elevates the issue beyond environment ministries.

References

Axelrod, Robert, and Robert Keohane. 1985. "Achieving Cooperation under Anarchy: Strategies and Institutions." *World Politics* 38 (October): 226–54.

Barrett, Scott. 1992. *International Agreements for the Protection of Environmental and Agricultural Resources: An Economics Perspective*. London Business School.

———. 1997. "The Strategy of Trade Sanctions in International Environmental Agreements." *Resource and Energy Economics* 19: 345–61.

———. 2003. *Environmental and Statecraft: The Strategy of Environmental Treaty-Making*. Oxford University Press.

———. 2007. *Why Cooperate? The Incentive to Supply Global Public Goods*. Oxford University Press.

Bhagwati, Jagdish. 2008. *Termites in the Trading System: How Preferential Agreements Undermine Free Trade*. Oxford University Press.

Biermann, Frank. 2000. "The Case for a World Environmental Organization." *Environment* 42, no. 9: 23–31.

Biermann, Frank, and Steffen Bauer, eds. 2005. *A World Environment Organization: Solution or Threat for Effective International Environmental Governance?* Aldershot, U.K.: Ashgate.

Bradford, Colin I., Jr. 2007. "World Energy Needs, Climate Change and Governance." Brookings and Centre for International Governance Innovation (www.l20.org/publications/28_4n_Global-Energy-Council-_full.pdf [October 2008]).

———. 2008. "The United States and Summit Reform in a Transformational Era." In *Emerging Powers in Global Governance. Lessons from the Heiligendamm Process*, ed. Andrew F. Cooper and Agata Antkiewicz. Waterloo, Ont.: Centre for International Governance Innovation and Wilfrid Laurier University Press.

Bradford, Colin I., and Johannes F. Linn, eds. 2007. *Global Governance Reform: Breaking the Stalemate*. Brookings.

Chang, Howard. 1997. "Carrots, Sticks and International Externalities." *International Review of Law and Economics* 17, no. 3: 309–24.

Charnovitz, Steven. 1993. "The Environment versus Trade Rules: Defogging the Debate." *Environmental Law* 23: 511–17.

———. 1997. "A Critical Guide to the WTO's Report on Trade and the Environment." *Arizona Journal of International and Comparative Law* 14, no. 2: 341–78.

———. 2000. "Rethinking WTO Trade Sanctions." Paper presented at conference on the Political Economy of International Trade, University of Minnesota, September 15–16.

———. 2002. "A World Environment Organization." UN University Institute of Advanced Studies.

Chayes, A., and A. Chayes. 1995. *The New Sovereignty: Compliance with International Regulation Agreements*. Harvard University Press.

Drezner, Daniel W. 2006. *U.S. Trade Strategy: Free versus Fair*. New York: Council on Foreign Relations Press.

———. 2007a. *All Politics Is Global: Explaining International Regulatory Regimes*. Princeton University Press.

————. 2007b. "The New New World Order." *Foreign Affairs*, March–April, 34–46.

————. 2008. "Two Challenges to Institutionalism." In *Can the World Be Governed? Possibilities for Effective Multilateralism*, ed. Alan Alexandroff. Waterloo, Ont.: Centre for International Governance Innovation and Wilfrid Laurier University Press.

Dunoff, Jeffrey. 1994. "International Misfits: The GATT, the ICJ, and Trade-Environment Disputes." *Michigan Journal of International Law* 15: 1043–127.

Esty, Daniel C. 1993. "GATTing the Greens." *Foreign Affairs* 72, no. 5 (November–December): 132–36.

————. 1994. *Greening the GATT: Trade, Environment and the Future*. Washington: Institute for International Economics.

————. 1999. "Toward Optimal Environmental Governance." *New York University Law Review* 74, no. 6: 1495–574.

Evans, Alex, and David Steven. 2007. "Climate Change: The State of the Debate." Center for International Cooperation, New York University.

Finger, J. Michael, and Philip Schaler. 2000. "Implementation of Uruguay Round Commitment: The Development Challenge." *World Economy* 23, no. 4: 511–25.

Frankel, Jeffrey. 2004. "Kyoto and Geneva: Linkage of the Climate Change Regime and the Trade Regime." Paper presented at Broadening Climate Discussion: The Linkage of Climate Change to Other Policy Areas, conference sponsored by Fondazione Eni Enrico Mattei and Massachusetts Institute of Technology, Venice, June.

Frieden, Jeffrey. 2006. *Global Capitalism: Its Fall and Rise in the Twentieth Century*. New York: W. W. Norton.

Goldstein, Judith, Miles Kahler, Robert O. Keohane, and Anne-Marie Slaughter, eds. 2001. *Legalization and World Politics*. MIT Press.

Gruber, Lloyd. 2000. *Ruling the World: Power Politics and the Rise of Supranational Institutions*. Princeton University Press.

Hauer, Grant, and C. Ford Runge. 1999. "Trade-Environment Linkages in the Resolution of Transboundary Externalities." *World Economy* 22, no. 1: 25–39.

Hudec, Robert. 1993. *Enforcing International Trade Law: The Evolution of the Modern GATT Legal System*. Salem, N.H.: Butterworth.

Ikenberry, G. John. 2000. *After Victory: Institutions, Strategic Restraint, and the Rebuilding of Order after Major Wars*. Princeton University Press.

Juma, Calestous. 2000. "The Perils of Centralizing Global Environmental Governance." *Environment Matters*.

Kaul, Inge, Isabelle Grunberg, and Marc A. Stern. 1999. *Global Public Goods: International Cooperation for the 21st Century*. Oxford University Press.

Kennan, George F. 1970. "To Prevent a World Wasteland: A Proposal." *Foreign Affairs* 48: 409–12.

Krasner, Stephen D. 1976. "State Power and the Structure of Foreign Trade." *World Politics* 28 (April): 317–47.

March, James, and Johan Olsen. 1998. "The Institutional Dynamics of International Political Orders." *International Organization* 52: 943–69.

Mumme, Stephen P. 1992. "New Directions in United States–Mexican Transboundary Environmental Management: A Critique of Current Proposals." *Natural Resources Journal* 31: 539–62.

North, Douglass. 1990. *Institutions, Institutional Change, and Economic Performance.* Cambridge University Press.

Palmer, Geoffrey. 1992. "New Ways to Make International Environmental Law." *American Journal of International Law* 86: 259–83.

Raustiala, Kal, and David Victor. 2004. "The Regime Complex for Plant Genetic Resources." *International Organization* 58: 277–309.

Richardson, Elliot L. 1992. "Climate Change: Problems of Law-Making." In *The International Politics of the Environment : Actors, Interests, and Institutions*, ed. Andrew Hurrell and Benedict Kingsbury. Oxford University Press.

Ricupero, Rubens. 1998. "UN Reform: Balancing the WTO with a Proposed 'World Environment Organization.'" In *Policing the Global Economy: Why, How, and for Whom?* ed. Sadruddin Aga Khan. London: Cameron May.

———. 2000. "A Development Round: Converting Rhetoric into Substance." Background note to the Conference on Efficiency, Equity and Legitimacy: The Multilateral Trading System at the Millennium, John F. Kennedy School of Government, Harvard University, June 1–2.

Runge, C. Ford. 2001. "A Global Environment Organization (GEO) and the World Trading System." *Journal of World Trade* 35, no. 4: 399–426.

———. 2009. "Multilateral Environmental Agreements (MEA's): Which Way Forward?" In *The Princeton Encyclopedia of the World Economy*, ed. K. Reinert and R. Rajan. Princeton University Press.

Runge, C. Ford, François Orlalo-Magné, and Philip Van de Kamp. 1994. *Freer Trade, Protected Environment: Balancing Trade Liberalization and Environmental Interests.* New York: Council on Foreign Relations Press.

Runge, C. Ford, and Benjamin Senauer. 2000. "A Removable Feast." *Foreign Affairs*, May–June, 39–51.

Sampson, Gary P. 2000. *Trade, Environment and the WTO: The Post-Seattle Agenda.* Policy Essay 27. Washington: Overseas Development Council.

Sandler, Todd. 1997. *Global Challenges: An Approach to Environmental, Political and Economic Problems.* Cambridge University Press.

Shaffer, Gregory C. 2001a. "The World Trade Organization under Challenge: Democracy and the Law and Politics of the WTO's Treatment of Trade and Environment Matters." *Harvard Environmental Law Journal* 25, no. 1.

———. 2001b. "WTO Blue-Green Blues: The Impact of U.S. Domestic Politics on Trade-Labor, Trade-Environment Linkages for the WTO's Future." *Fordham International Law Review* 24: 608–51.

Slaughter, Anne-Marie. 2004. *A New World Order: Government Networks and the Disaggregated State.* Princeton University Press.

Snidal, Duncan. 1985. "The Limits of Hegemonic Stability Theory." *International Organization* 39: 579–614.

Susskind, Lawrence, and Connie Ozawa. 1992. "Negotiating More Effective International Environmental Agreements." In *The International Politics of the Environment : Actors, Interests, and Institutions*, ed. Andrew Hurrell and Benedict Kingsbury. Oxford University Press.

Thacher, Peter S. 1992. "The Role of the United Nations." In *The International Politics of the Environment : Actors, Interests, and Institutions*, ed. Andrew Hurrell and Benedict Kingsbury. Oxford University Press.

UN Task Force on Environment and Human Settlements. 1998. *Recommendations*. Document A/53/463. New York: United Nations.

von Moltke, Konrad. 2001. "Wither MEAs? The Role of International Environmental Management in the Trade and Environment Agenda." International Institute for Sustainable Development.

Whalley, John, and Ben Zissimos. 2000. "A World Environmental Organization?" University of Western Ontario, University of Warwick, and National Bureau of Economic Research.

Reflections on Climate Change and Trade

Let me start with a general comment that is relevant as background to the theme of this book, and then move on to some of the specifics of the interface between trade, the World Trade Organization, and the environment that many of the chapters above have addressed. At the outset, we need to remember that those who work on trade (mostly academics) and those who work on the environment (mostly activists) have traditionally been at loggerheads from time to time.

Why? One important philosophical difference that underlies much of this tension, which I think we tend to forget, is that trade economists are typically considering and condemning governmental interventions (specifically, protectionism, such as the imposition of tariffs and nontariff barriers) mainly as *creating* distortions and harming the general welfare. Conversely, environmentalists are typically dealing with what are best described as "missing markets" (for example, people dump carcinogens into lakes, rivers, and oceans and emit them into the atmosphere, and they do not have to pay for the pollution). Therefore, they see government intervention (for example, the use of pollution-pay taxes or the use of tradable permits) as *correcting* a distortion. It is useful to recall this fundamental difference in the experiences and lifestyles of the people on the two sides of the trade-and-environment aisle, because it underlies and explains, to some extent, the occasional frictions between them.

Of course, trade and the environment are integrally related, and that is why many disputes were coming up at the General Agreement on Tariffs and Trade (GATT)—the most important being, of course, the celebrated dolphin-tuna case between the United States and Mexico. I will return to the important issues raised by the dolphin-tuna jurisprudence and its later reversal in the shrimp-turtle dispute, also involving the United States. But let us start with the problems raised by global warming.

In his comment on chapters 5 and 6, Daniel Drezner points out that, in the past, America has opted for short-run adjustment costs with a view to long-run gain. I do not quite know what he means by "short-run adjustment costs," but

I would simply say that an enlightened hegemon like the United States, when going in for the GATT, certainly did not insist on the developing countries having reciprocal obligations. It simply gave away membership. It was, in fact, getting the developing countries into the GATT while gaining nothing in terms of an immediate, reciprocal opening of markets.

I think the intention was to create more legitimacy for the GATT by increasing membership. Down the road, then, you would have graduation and begin to "collect"—via what is called extended reciprocity or intergenerational reciprocity. There were no short-run costs à la Drezner either. After all, the developing countries at that time, in the mid-1940s, were not important markets anyway; nor were they, by and large, major exporters. They were really small players in world trade, and it was only later, when they had grown, that the usual question of reciprocity would become economically relevant.

So one may ask why this argument does not work with the Kyoto Protocol at the moment. Why are we not willing to play that old GATT game? I think we need to look into this question carefully to get a sense of where the problems might be with how we approach the design of the successor pact to Kyoto—what I like to call Kyoto II.

A key problem, of course, is that there are two big players, India and China, with current and prospective emissions of carbon dioxide that are simply large for India and huge for China. We did not have anything like this at the time the GATT was formed; then, as noted, the developing countries were all little players in trade, for all practical purposes. Exempting India and China from the emission obligations of a climate change treaty today is thus not like exempting the developing countries from trade obligations in the 1940s. Moreover, India and China are not willing to make any payment to get into the Kyoto club, as it were, simply because they feel—and this is where, I think, the real crux comes in—that they did not contribute to past environmental damages.

Now, if one looks at the past environmental damages, it is clear that the accumulated fossil fuel carbon dioxide for 1850 to 2005 shows the damage attributable to India and China is about 10 percent, whereas the countries now belonging to the European Union, Russia, and the United States jointly account for over 60 percent. So you have basically what I have called a "stock" problem,[1] the problem of "past" damage to the environment—for which America and the EU, basically, are particularly responsible. And the solution to this "stock" problem in the Kyoto Protocol, which was devised to bring India and China on board, was to say, "Look, because we were the ones who imposed large losses on the environment in the past, and not you, we will exempt you from any 'flow' obligation for reducing the current damage, no matter how large."

1. This was in a center-page *Financial Times* op-ed article in August 2006.

Now, the problem is that, in so designing the Kyoto Protocol, its framers were trying to kill two birds with one stone. And, of course, that stone is not something palatable—to mix metaphors—to the U.S. Congress. The U.S. Senate virtually unanimously rejected Kyoto in 1997 because its members thought that India and China were going to be free riders, when in fact the free ride was being provided because they had not been riding for almost a hundred years while America had been! I think that the general feeling instead was that these countries were being let off simply because they were developing countries, presumably on a progressive taxation ground; but progressive taxation has become increasingly a hard sell (though the Barack Obama administration may well restore it to some respectability).

In sum, India and China were *not* free riders. Rather, their governments were saying to the Western nations: "Look, you have done a lot of damage. You've got this 'stock' liability for past emissions. And you cannot just get us to accept significant 'flow' liability for current emissions while you do nothing significant on the stock side."

Thus, I have always felt that the Kyoto Protocol was doomed, in a way, because it really could not be effective, as designed, until we addressed this particular basic issue clearly and directly in a transparent manner—forgoing the fudge that mixed up the stock and the flow dimensions of the obligations. And I think this problem is going to afflict Kyoto II as well. Frankly, what we are negotiating so far shows little willingness on the part of today's rich, developed countries to accept the notion that they must pay for past damages; and so it would be little short of ethical nonsense for them to ask India and China to accept much larger flow obligations.

Now, this unwillingness to face up to the liability for past carbon emissions is rather strange, in the sense that the United States has already accepted,[2] in its domestic environmental practice, the superfund approach under which, for hazardous waste, liability has been assigned, in eligible industries, for past damages, even when the pollution was not regarded at the time to be harmful. America is a nation that thrives on torts; indeed the Democratic Party does also, and it cannot be denied that America has actually accepted the superfund approach in its own environmental policy.

So, in my judgment, for us to go around saying that India and China have to accept obligations on the flow side—which I think is perfectly appropriate— while doing nothing like a substantial superfund for past carbon emissions on the stock side, is to invite condemnation as a superpower play by nations, both the EU members and the United States, that are no longer quite the superpow-

---

2. I noted this in my 2006 *Financial Times* article.

ers that they were once. You really need to walk on two legs and not just on one leg.

I see statements all the time, from even Al Gore and Bill Clinton, about the desirability of China and India accepting flow obligations. But unless I have missed something, neither has publicly acknowledged the need for a substantial superfund for the U.S. stock liability. So much for their environmentalism: self-serving for the United States, not cosmopolitan and just.

Now, the same problem arises in trade negotiations because India, and several developing nations, say to America, "How can you have to this day sizable trade-distorting agricultural subsidies, and then expect us to open our agricultural sector to competition from such subsidized exports by you?" In fact, the Doha Round multilateral trade talks collapsed in August 2008 precisely because India claimed that nearly two-thirds of its people were in the farming sector, most were subsistence farmers, and the United States had only 2 million, often large, farmers with much larger subsidy support. So, India wanted a special safeguards mechanism that, in my view, was excessively cautious, citing our subsidies. Remember, of course, that the United States itself had introduced special safeguards against China; and that nothing works better to get protection than to allege, often without any basis, that the exporters are "unfairly" subsidizing exports to us. Yet, when the talks collapsed, the U.S. trade representative and an obliging media, and Congress in turn, zeroed in on India as the rejectionist culprit.

So, as one draws analogies between trade and the environment, it is necessary to remember that unless America brings to both negotiations, each of which is extremely important, the notion that it cannot just impose what it wants on others, often to its presumed advantage regardless of the others', it is likely to meet with failure. Charles Kindleberger famously called the United States an "altruistic hegemon." I fear that it has increasingly tended in recent years to become a selfish hegemon.

I should add that it is not just the United States that is a problem. I see little attention being paid to the stock problem in Europe either. As then–senator Obama said about Senator John McCain: He is a good man; it is just that he does not get it. Thus, when I was in Florence recently, and Tony Blair was in the chair and talking about what he was doing on the environment, Kishore Mahbubani, myself, and others from the developing countries drew his attention to the superfund idea. He continued through the session as if he had heard nothing. As then–senator Obama would have said: Prime Minister Blair just does not get it. But unless he, Gore, Clinton, and others do get it, do not expect that Kyoto II can be signed and ratified by India, China, the United States, and European Union.

Now, let us turn to the problem raised by the notion—fashionable in Congress these days—that if India and China do not accept green house emissions obligations, America would impose a "border tax," better called an import duty, that is equal to the carbon tax that they are not imposing in sync with America's. This is, of course, like the idea that the French floated against the United States, saying they would tax American exports to the EU because America had not signed the Kyoto Protocol. This issue, of course, takes us back to the tuna-dolphin case in 1991 at the GATT. When tuna-dolphin came up, the environmentalists were terribly upset that the United States lost the case. At that time, I happened to be the economic policy adviser to Arthur Dunkel, the director-general of GATT. And so I was consulted by the legal adviser, Frieder Roessler, on the ongoing case and what the position of the GATT Secretariat should be. The focus at the time was whether specific process and production methods (PPM) should be allowed to be prescribed for import eligibility—that is, could the United States specify that tuna should be allowed to be imported only if purse-seine nets that also caught dolphins were not used?

Coming from the economic side, I felt that PPMs, as a general case, should not be allowed to be so used to regulate the entry of imports because they could thus be used to discriminate against specific suppliers while appearing to be nondiscriminatory. After all, those involved in international trade have all been brought up on the famous apocryphal example (based, however, on a real case) of imported cheese being taxed by Germany if it was produced by cows grazing at 4,500 feet and above, with bells around them and under Alpine conditions. This was obviously aimed at Swiss cheese, although, in principle, if Tanzania were to satisfy the conditions, Tanzanian cheese would be equally subjected to the same high tariff. The use of the PPM could then defeat the intent of nondiscrimination required by the GATT.

So we were coming at the PPM issue from the trade side, because the GATT was a trade institution. And we did not really think of the environmental aspect specifically at that particular point (except that, if the issue fell under Article XX, greater leeway was permissible).

Thus the position we took was that the legitimation of a free use of PPMs to regulate imports would open the door to the indiscriminate use of de facto discrimination in trade among different suppliers, undercutting the basic principle of nondiscrimination underlying the GATT. Anybody could say the way you produce something, no matter how or why, is unacceptable. We could not see how de facto discrimination could be contained; it could proliferate hugely.

But the shrimp-turtle decision years later, by the Appellate Body of the World Trade Organization, basically reversed the dolphin-tuna jurisprudence, ignoring our caution. It meant that we would now be opening the floodgates for all kinds of PPM prescriptions that would afflict anyone, on any issue (though we had also argued that the situation would be asymmetrical between weaker and stronger nations because it was unlikely that the weaker nations could take on the stronger nations in this essentially arbitrary fashion—a worry that has also been expressed by prominent nongovernmental organizations in the developing countries).

I was among the few who thought that this decision was ill judged, revealing the weakness of an Appellate Body where familiarity with legal jurisprudence and practice is not a requirement for an appointment. Now, I would simply say that the chickens have come home to roost against the United States itself. The French plan to tax imports from the United States because the United States had not signed on to the Kyoto Protocol was exactly the kind of thing I had predicted. And now the United States, which has among the lowest gasoline prices in the world, absurdly believes that, instead of being subjected to PPM restrictions itself on grounds of inadequate energy prices, it can put import taxes on such PPM grounds against India and China.

And, frankly, what would then prevent India from discriminating against U.S. exports on the ground that the United States does not have a superfund? I could go on endlessly. This way lies chaos, just as I had argued to Roessler, and to Dunkel, during the dolphin-tuna panel's deliberations.

I think America needs to be very careful about not going down the route that has been opened up by the U.S. legislation and the World Trade Organization's ruling in support of it in the shrimp-turtle case. If America goes down that legislative route, it is likely to be the loser in the end—certainly on energy and the environment. Thus, Congress needs to be told that this is a game everybody can play.

Contributors

JOSEPH E. ALDY is a fellow at Resources for the Future and codirector of the Harvard University Project on International Climate Agreements. His research focuses on climate policy, mortality risk valuation, and energy policy. He is also codirector of the International Energy Workshop and an adjunct assistant professor at Georgetown University. From 1997 to 2000, he served on the staff of the President's Council of Economic Advisers, where he was responsible for domestic and international climate change policy. He participated in bilateral and multilateral conferences and meetings on climate policy in Argentina, Bolivia, China, France, Germany, Kazakhstan, South Korea, Israel, Mexico, and Uzbekistan, including the Conference of Parties (COP)-4, COP-5, the Organization for Economic Cooperation and Development, and the International Energy Agency. He coedited (with Robert Stavins) *Architectures for Agreement: Addressing Global Climate Change in the Post-Kyoto World* (Cambridge University Press, 2007).

WILLIAM ANTHOLIS is the managing director of Brookings. He has been a senior adviser on foreign security and economic policy at the National Security Council and the State Department, and was director of studies at the German Marshall Fund of the United States, a U.S. grant-making and public policy institution devoted to strengthening transatlantic cooperation. He has been an international affairs fellow of the Council on Foreign Relations and a visiting fellow at the Center of International Studies at Princeton University. In 1991, he cofounded the Civic Education Project, a nonprofit organization that supports Western-trained social science instructors at universities in nineteen Central and Eastern European countries.

JAGDISH BHAGWATI is university professor at Columbia University and a senior fellow in International Economics at the Council on Foreign Relations. He has been economic policy adviser to Arthur Dunkel, director-general of the General Agreement on Tariffs and Trade (1991–93), special adviser to the United

177

Nations on globalization, and external adviser to the World Trade Organization, where he served on the expert group on the organization's future. He was on the advisory committee to UN secretary-general Kofi Annan for the New Partnership for Africa's Development process, and was also a member of the eminent persons' group under the chairmanship of President Fernando Henrique Cardoso on the future of the United Nations Conference on Trade and Development. Five volumes of his scientific writings and two of his public policy essays have been published by the MIT Press. He has been the recipient of six festschrifts, and he has also received several prizes and honorary degrees, including awards from the governments of India (Padma Vibhushan) and Japan (the Order of the Rising Sun and the Gold and Silver Star). His latest book, *In Defense of Globalization* (Oxford University Press, 2004), was published to worldwide acclaim.

JASON E. BORDOFF is policy director of the Hamilton Project, an economic policy initiative housed at the Brookings Institution committed to promoting more broadly shared prosperity. He has written on a broad range of economic policy matters, with a focus on climate and energy, trade and globalization, and tax policy. His publications include *Path to Prosperity* (Brookings Institution Press 2008, with Jason Furman) and several journal articles, book chapters, op eds and policy papers. He is also a term member of the Council on Foreign Relations and serves on the board of the Association of Marshall Scholars. He is a member of the New York and Washington, D.C., bar associations. He previously served as an adviser to Deputy Secretary Stuart E. Eizenstat at the U.S. Treasury Department, and worked as a consultant for McKinsey & Co. in New York. He graduated with honors from Harvard Law School, where he was treasurer and an editor of the Harvard Law Review, and clerked on the U.S. Court of Appeals for the D.C. Circuit. He also holds an MLitt degree from Oxford University, where he studied as a Marshall Scholar, and a BA from Brown University.

NILS AXEL BRAATHEN is a principal administrator in the Environment Directorate of the Organization for Economic Cooperation and Development (OECD), where he has worked since 1996. He is currently working on environmentally related taxes, the valuation of environmental externalities, the environmental effects of transport, and a database on instruments used for environmental policy. He has previously done macroeconomic model simulations of environmental policies and researched voluntary approaches to environmental policy, the economics of waste, and instrument mixes for environmental policy. He holds a master's degree in economics from the University of Oslo. Before joining the OECD, he was deputy director-general in the Department for Long-Term Planning and Policy Analysis of the Norwegian Ministry of Finance and deputy director-general in the Industrial Policy Department of the Norwegian Ministry of Industry. From 1987 to 1989, he was deputy director of the Association of Norwegian Finance Houses, and from 1981 to 1987, an administrator and then division head in the Norwegian Ministry of Industry.

COLIN I. BRADFORD JR. is a nonresident senior fellow at Brookings and the Centre for International Governance Innovation (CIGI) in Waterloo, Canada. He was previously chief economist at the U.S. Agency for International Development, head of research at the Development Center of the Organization for Economic Cooperation and Development, a senior staff member of the Strategic Planning Unit of the World Bank, research professor of economics and international relations at American University, and associate professor of international economics and management at Yale University. His research and expertise are focused on global poverty, globalization, foreign assistance, international economics, and international organizations.

LAEL BRAINARD is vice president and founding director of Brookings Global Economy and Development and holds the Bernard L. Schwartz Chair in International Economics at Brookings. She served as the deputy national economic adviser and chair of the Deputy Secretaries Committee on International Economics during the Bill Clinton administration. As the deputy director of the National Economic Council, she helped build a new White House organization to address global economic challenges such as the Asian financial crisis and China's entry into the World Trade Organization. Previously, she served as an associate professor of applied economics in the Sloan School of Management at the Massachusetts Institute of Technology. She is currently a member of the Wesleyan University Board of Trustees, the Council on Foreign Relations, and the Aspen Strategy Group.

THOMAS L. BREWER is on the faculty of the McDonough School of Business at Georgetown University and research director of Climate Strategies. He specializes in issues associated with climate change, including the intersections of climate change issues with international trade, technology transfer, and investment issues. He has written or edited numerous books and published articles in a wide range of refereed journals on international business issues, including trade and investment issues at the World Trade Organization. He has made presentations on climate change issues at the Royal Institute of International Affairs (Chatham House), the European Commission, the European Parliament, the Centre for European Policy Studies in Brussels, the Economic and Social Research Institute of the Japanese government, the World Bank, and the International Centre for Trade and Sustainable Development, as well other conferences for academic, business, and government audiences.

DANIEL W. DREZNER is professor of international politics at the Fletcher School of Tufts University. He has previously taught at the University of Chicago and the University of Colorado at Boulder. He is the author of *All Politics Is Global: Explaining International Regulatory Regimes* (Princeton University Press, 2007), *U.S. Trade Policy: Free versus Fair* (Council on Foreign Relations Press, 2006), and *The Sanctions Paradox: Economic Statecraft and International Relations* (Cambridge University Press, 1999). He has published articles in numerous scholarly journals as well the *New York Times*, *Wall Street Journal*, *Washington Post*, and *Foreign Affairs*. He has received fellowships from the German Marshall Fund of the United States, the Council on Foreign Relations, and Harvard University, and he previously held positions with the Civic Education Project, the RAND Corporation, and the U.S. Treasury Department. He is a regular commentator for *Newsweek International* and National Public Radio's *Marketplace*. He keeps a daily weblog for *Foreign Policy* (www.drezner.foreignpolicy.com). His next book, *An Unclean Slate*, will examine the future of global governance.

JEFFREY A. FRANKEL is Harpel Professor of Capital Formation and Growth at Harvard University's John F. Kennedy School of Government. He directs the program in International Finance and Macroeconomics at the National Bureau of Economic Research, where he is also a member of the Business Cycle Dating Committee, which officially declares recessions. He was appointed to the Council of Economic Advisers by President Bill Clinton in 1996 and served until 1999, with responsibility for international economics, macroeconomics, and the environment. He was previously a professor of economics at the University of California, Berkeley, having joined the faculty in 1979. His other past appointments include the Federal Reserve Board, Institute for International Economics, International Monetary Fund, and Yale University. His research interests include international finance, currencies, monetary and fiscal policy, commodity prices, regional blocs, and global environmental issues.

(TOM) HU TAO is professor of Renmin University of China and an affiliated faculty member of College of Business, Oregon State University. He was previously the chief economist for the Policy Research Center for Environment and Economy in the Ministry of Environmental Protection of the People's Republic of

China. He served as the Ministry's chief expert on trade and the environment at the Doha Round of World Trade Organization negotiations, and he was also a member of the China Council for International Cooperation on Environment and Development. With educational experience in ecology, environmental economics, and agricultural economics, he specializes in issues related to international environmental policy, international trade, and globalization and climate change. He has worked on many projects with organizations such as the Asian Development Bank, World Bank, United Nations Development Program, Global Environment Facility, International Institute for Sustainable Development, and Norwegian Ministry of the Environment.

ARIK LEVINSON is an associate professor in the Economics Department of Georgetown University, where he teaches microeconomics, public finance, and environmental economics. Most of his research involves the fields of traditional public finance and environmental economics. His current projects concern explanations for the reductions in pollution from U.S. manufacturing, valuations of air quality using happiness data, environmental Engel curves, and the backward-bending quality supply curve of foster parents. He is also a research associate at the National Bureau of Economic Research and a coeditor of the *Journal of Environmental Economics and Management*.

MUTHUKUMARA MANI is a senior environmental economist in the World Bank. His work has focused on trade and climate change, pollution prevention policy, natural resources management, environmental taxes, environmental institutions and governance, and trade and environment issues. His research has appeared in numerous peer-reviewed journals in the fields of economics, the environment, and political economy, and he has also coauthored several policy research working papers for the World Bank and the International Monetary Fund. Before joining the World Bank, he was an economist in the Fiscal Affairs Department of the International Monetary Fund, where he was responsible for analyzing the environmental implications of macroeconomic policies and programs and for integrating broad environmental considerations in country programs.

WARWICK J. MCKIBBIN is a nonresident senior fellow at Brookings and director of the Centre for Applied Macroeconomic Analysis in the College of Business and Economics at the Australian National University. He is also a professorial fellow at the Lowy Institute for International Policy in Sydney and president of McKibbin Software Group. He is a member of the Board of the Reserve Bank of Australia and a member of the Australian Prime Minister's Science, Engineering, and Innovation Council. He recently served as a member of the Australian Prime Minister's Taskforce on Uranium Mining Processing and Nuclear Energy. He has been a consultant for many international agencies and governments on issues of macroeconomic policy, international trade and finance, greenhouse policy, global demographic change, and the economic cost of infectious diseases.

C. FORD RUNGE is Distinguished McKnight University Professor of Applied Economics and Law and former director of the Center for International Food and Agricultural Policy at the University of Minnesota. He has served on the staff of the U.S. House of Representatives Committee on Agriculture. As a Science and Diplomacy Fellow of the American Association for the Advancement of Science, he worked at the U.S. Agency for International Development on food aid and trade. He spent 1988 as a special assistant to the U.S. ambassador to the General Agreement on Tariffs and Trade in Geneva. In 1988, he was named a member of the Council on Foreign Relations in New York, and in 1990 a Fulbright Scholar for study in Western Europe. His publications include five books and a wide range of articles concentrating on trade, agriculture, and natural resources policy, including *Freer Trade, Protected Environment: Balancing Trade Liberalization and Environmental Interests* (Council on Foreign Relations Press, 1994) and *Ending Hunger in Our Lifetime: Food Security and Globalization* (International Food Policy Research Institute, 2003). His article "How Biofuels May Starve the Poor" appeared in the May–June 2007 *Foreign Affairs*.

ANDREW W. SHOYER is a partner in the Washington office of Sidley Austin LLP and chairs the firm's international trade and dispute resolution practice. He focuses on the implementation and enforcement of international trade and investment agreements. Drawing on his experience at the Office of the U.S. Trade Representative (USTR) and the World Trade Organization (WTO), he advises companies, trade associations, and governments on the use of WTO and other treaty-based trade and investment rules to open markets and resolve disputes. He spent seven years at the USTR, serving most recently as legal adviser in the U.S. Mission to the WTO in Geneva. He was the principal negotiator for the United States of the rules implementing the WTO Dispute Settlement Understanding, and he has briefed and argued numerous WTO cases before dispute settlement panels and the WTO Appellate Body. He is also an adjunct faculty member in international trade policy at the School of Foreign Service of Georgetown University.

ISAAC SORKIN is a research assistant for Brookings Global Economy and Development.

PETER J. WILCOXEN is a nonresident senior fellow at Brookings, an associate professor of economics and public administration at the Maxwell School of Syracuse University, and the director of the Maxwell School's Center for Environmental Policy and Administration. His principal area of research is the effect of environmental and energy policies on economic growth, international trade, and the performance of individual industries. He is a coauthor (with Dale W. Jorgenson) of the Jorgenson-Wilcoxen Model, a thirty-five-sector econometric general equilibrium model of the U.S. economy. He is also a coauthor (with Warwick J. McKibbin) of G-Cubed, an eight-region, twelve-sector general equilibrium model of the world economy that has been used to study international trade and environmental policies. He is currently a member of the Environmental Economics Advisory Committee of the U.S. Environmental Protection Agency's Science Advisory Board, and he was a review editor on the *Third Assessment Report of the Intergovernmental Panel on Climate Change* (Cambridge University Press, 2001).

Index

Page numbers followed by the letters f and t refer to figures and tables, respectively.

185

www.ingramcontent.com/pod-product-compliance
Lightning Source LLC
Chambersburg PA
CBHW021816270326
41932CB00007B/213

9 780815 702986